VITAMIN C

A 500-YEAR SCIENTIFIC BIOGRAPHY FROM SCURVY TO PSEUDOSCIENCE

Stephen M. Sagar

Prometheus Books

Essex, Connecticut

Ⓟ Prometheus Books

An imprint of Globe Pequot, the trade division of
The Rowman & Littlefield Publishing Group, Inc.
4501 Forbes Blvd., Ste. 200
Lanham, MD 20706
www.rowman.com

Distributed by NATIONAL BOOK NETWORK

British Library Cataloguing in Publication Information Available

Library of Congress Cataloging-in-Publication Data

Names: Sagar, Stephen M., author.
Title: Vitamin C : a 500-year scientific biography from scurvy to
    pseudoscience / Stephen M. Sagar.
Description: Lanham, MD : Rowman & Littlefield, [2022] | Includes
    bibliographical references and index. | Summary: "Vitamin C: A 500-Year
    Scientific Biography from Scurvy to Pseudoscience is the compelling
    story of the history and science behind vitamin C"— Provided by
    publisher.
Identifiers: LCCN 2021060986 (print) | LCCN 2021060987 (ebook) | ISBN
    9781633888265 (cloth) | ISBN 9781633888272 (epub)
Subjects: LCSH: Vitamin C—History.
Classification: LCC QP772.A8 S24 2022 (print) | LCC QP772.A8 (ebook) |
    DDC 615.3/28—dc23/eng/20211220
LC record available at https://lccn.loc.gov/2021060986
LC ebook record available at https://lccn.loc.gov/2021060987

For Susan

[R]esearch is not a systematic occupation but an intuitive artistic vocation.

—Albert Szent-Gyorgyi,
"Lost in the Twentieth Century," 1963

I don't think anything is ever just science.

—Siri Hustvedt, *What I Loved*, 2003

CONTENTS

PREFACE

I first encountered pure ascorbic acid, the chemical name for vitamin C, in 1977. Having just completed medical school and residencies in internal medicine and neurology, I aspired to be a clinician-scientist and went to work in a neuroscience laboratory in Boston. In that job, I used ascorbic acid virtually every day. But I was not interested in the compound; it was merely a substance to prevent the oxidation of other reagents. I did not bother to weigh it out, adding the few milligrams picked up on a spatula tip.

Vitamin C first piqued my interest in the early 1980s when it earned me funding for a pilot research project. One of the Sackler brothers invited my mentor to New York to discuss his laboratory's work. The Sackler family is now infamous for owning Purdue Pharma, promoting OxyContin, and, in the minds of many, igniting the opioid crisis. But that was before OxyContin. The family had gotten rich advising big pharmaceutical companies how to promote drugs directly to doctors. They devoted some of their wealth to a foundation that supported medical research, and my mentor asked his trainees for brief grant proposals to take to his meeting to try to capture some of that wealth for his laboratory.

My proposal concerned the neurobiological actions of capsaicin and, like most writings about capsaicin, contained a statement that it is the active, pungent ingredient of Hungarian paprika and Mexican chili peppers. The Sackler brother who perused the proposal was not interested in capsaicin or in my science. He decided to fund the grant because Hungarian paprika had been a commercial source of vitamin C. Unbeknownst to me, his foundation was supporting the research of Linus Pauling, the Nobel Prize–winning chemist and most publicly visible proponent of megadose vitamin C. My grant proposal had nothing to do with vitamins and nothing of importance came of the project, but this small windfall was one example of the serendipity that punctuated the history of vitamin C.

My research went in other directions and I had little more to do with vitamin C until I switched tracks and entered the pharmaceutical industry. In 2012, I prepared a talk for a job interview at a small biotech company. The same mechanism that transports one form of vitamin C into the brain also transports the drug that I would have worked with at this company. This led me to investigate the neurological manifestations of scurvy and delve into the history of that disease.

I did not take the job, but as I read more, the intriguing stories and fascinating characters in the history of vitamin C convinced me that it was a story worth bringing to a wider audience. The five-hundred-year saga is rich with compelling stories and interesting characters, with tales of courage and callousness, of brilliant insight and folly, and of strokes of luck, both good and bad. The protagonists include flamboyant personalities with conflicting egos—swashbuckling sailors, Arctic explorers, penny-pinching bureaucrats, scientists working in malaria-infested jungle laboratories, and investigators utilizing the latest tools of molecular biology. The story provides examples of our failure to learn from history and to repeat the mistakes made by our ancestors hundreds of years ago.

The history of vitamin C illustrates how medical science works in the real world and how it has changed over the centuries. It shows that the human brain can penetrate the mysteries of biology but also can be led astray. It illuminates the fitful progression of science as scientists struggle against the limitations of human intellect while society attempts to deal with the implications of those discoveries. The COVID-19 pandemic convinced me even more of the importance of looking to the history of science to better understand our current dilemmas.

INTRODUCTION

*Miguel: I thought it might have been one of those
made-up things that adults tell kids, like vitamins.*

Aunt Victoria: Miguel, vitamins are a real thing.

—*Coco*, Pixar Animation Studios, 2017

A SPECIAL KIND OF MALNUTRITION

Vitamin C is an essential nutrient for humans. It allows us to
live in an atmosphere rich in oxygen and to use that oxygen
to fuel our bodies without burning up our tissues in the process.
The ability to make vitamin C evolved in animals when the first
amphibians ventured onto land four hundred million years ago and
encountered an atmosphere with forty times the concentration of
oxygen as the ocean. Vitamin C enabled the expansion of the animal kingdom over the earth. However, apes, living in forests with
fruits and plants rich in vitamin C, lost the ability to make their own.
Their diets provided more than enough.

The first humans inherited this deficiency from their ape ancestors and depended on food for their vitamin C. This was not a problem for hunter-gatherers or settled agrarian populations eating a diversified diet; many plants make vitamin C in abundance. Fresh fruits, berries, green vegetables, potatoes, peppers, and cabbage are rich sources of the vitamin (see appendix).

When humans migrated out of the African forest and into Europe and Asia, foods providing vitamin C were not always available. People living in northern latitudes during winter, armies with stretched supply lines, and sailors on long ocean voyages had to go without fresh fruits and vegetables for months at a time. The disease scurvy, caused by vitamin C deficiency, became common. Scurvy occurred during ancient times and throughout the Middle Ages, but the first clear description of the disease appeared in the journal of Vasco da Gama's first voyage to India at the end of the fifteenth century.

Having eaten no fresh fruit or vegetables for more than six months, da Gama and his men sailed up the east coast of Africa and began to fall ill with a disease for which they had no name. Their limbs swelled, their gums grew over their teeth, and they stopped eating. They became too debilitated to man the ship, and many went on to ugly and painful deaths.

Da Gama discovered the cure. Encountering orange trees on the African coast, the men avidly ate the fruit and recovered. Despite this knowledge, as long ocean voyages became frequent during the Age of Sail, scurvy became the greatest danger faced by mariners, exceeding the risk of dying in storms or battles. Scurvy also struck those on land, and millions of Europeans died of the disease.

How could so many have died from a curable illness? In essence, it was because doctors and bureaucrats had no concept of vitamins and held outmoded ideas about disease. It took four hundred years to understand the simple fact that scurvy resulted from a nutritional deficiency.

Early in the twentieth century, the nature of scurvy was finally recognized, and chemists went to work to identify the curative nutrient. In the early 1930s, scientists isolated the substance that cured scurvy and determined its chemical structure. It was named *ascorbic acid* because it is the antiscorbutic (meaning "anti-scurvy") substance. Factories produced it in bulk, and laboratories embarked on an ongoing effort to define its role in our bodies. Developments in transportation and agriculture made fresh produce available throughout the world and throughout the year. Ascorbic acid was added to foods and beverages as a preservative. Scurvy all but disappeared from the developed world.

However, that is not the end of the story. Although a single orange contains more than five times the amount of vitamin C necessary to prevent scurvy, the two-time Nobel Prize–winning chemist Linus Pauling proclaimed that humans require *grams* of vitamin C each day to keep healthy, not just the few milligrams needed to prevent scurvy. Other experts disagreed and pursued an answer to the question, *how much is enough?* Do we need to ingest just enough to prevent scurvy, or does ascorbic acid at high doses have other benefits? In the 1970s, these questions escaped the realm of medical science and entered politics and commerce. Vitamins became big business.

VITAMIN C MAKES MONEY

During the past fifty years, vitamin C has been the world's favorite vitamin, outselling all others. In 2016, it slipped into second place behind vitamin D in the United States, but sales continued to grow. It roared back into first place in 2020 with the COVID-19 pandemic.

Factories, mainly in China and India, produce more than 150,000 tons of vitamin C every year, worth a billion dollars. Consumers buy about one-third as a nutritional supplement. Much of the rest shows up in food and beverages and in animal feed as a preservative.[1] Amazon sells more than a hundred vitamin C products, including tablets, capsules, chewables, powders, liquids, drops, creams, and sprays. Vitamin C was the trailblazer for the evergrowing $40 billion vitamin and supplement industry. Although other products have grown to such an extent that it now accounts for only 2 percent of that market, health-conscious consumers continue to buy the vitamin.

ONE CONSUMER OF VITAMIN C

A friend whom I will call Mary is a healthy, active executive of a nonprofit organization. A few times each year when she feels she might be coming down with a cold, she takes a dissolvable powder that contains one thousand milligrams of vitamin C, along with modest amounts of vitamins A and E and some minerals and herbs. One thousand milligrams of vitamin C is 1,667 percent of the "daily value" according to the list of ingredients on the label, which says the mixture "supports the immune system" but makes no specific claims that it prevents or treats a cold or other infection.

Mary takes two or three doses over a day or two. Frequently she can "fight off the cold," but sometimes her symptoms progress to a full-blown upper respiratory infection despite taking the vitamin. She then stops taking it, as she does not believe it lessens her symptoms once the cold has set in.

Mary describes herself as a show-me-the-evidence kind of person. When asked why she takes vitamin C after clinical trials have shown that there is no benefit in taking it at the onset of cold symptoms, she replies that her experience has been different. What is her experience? She takes the pills a few times a year and usually has only one or two colds annually. She figures that the tablets work much more often than not, and she keeps taking them.

ANOTHER CONSUMER OF VITAMIN C

Another friend, Bill, is an eminent Harvard-educated physician-scientist and a professor at a leading medical school. One subject he studies is the role of oxygen in muscle damage. Several years ago when he was competing in marathons, he took five hundred-milligram vitamin C tablets daily, hoping to promote the health of his muscles. He did not think his diet was deficient in vitamin C, but he reasoned that his long-distance running subjected his muscles to an extreme level of stress, so he might benefit from antioxidant protection from extra vitamin C. He joined the millions of consumers who take one of the many formulations of vitamins, minerals, and herbs touted as antioxidants.

He thought, "Why not?" reasoning that the vitamin might do some good, but, as a natural product, he did not believe it would do any harm. He did not know how much of each dose was absorbed into his body nor how much merely disappeared down the toilet. At some point, for no reason that he can recall, he stopped taking the pills and did not notice any difference.

Mary and Bill are examples of the millions of people who take supplemental vitamin C. Linus Pauling and his disciples, along with the promotional material from the vitamin industry, have convinced many that they need to ingest grams of the vitamin to stave off disease. They believe that the typical Western diet is inadequate to maintain health and that it must be supplemented with large doses of industrially synthesized vitamins.

Millions once died of scurvy because of the lack of a few milligrams of vitamin C; now millions consume grams of vitamin C. The story of this progression occurs in three parts.

SCURVY: A LONG AND WINDING ROAD
TO UNDERSTANDING

We take vitamins for granted. We enjoy a wide variety of foods, many enriched with vitamins. As a result, we have forgotten the once-common diseases that occurred when people survived on restricted diets. The names sound strange: scurvy, beriberi, pellagra, rickets. Physicians know them only from textbooks; few ever see a case.

However, before the twentieth century, the question was not whether to take extra vitamins, but how to get enough to stay alive. Millions died from a lack of vitamins. In some countries, vitamin deficiencies killed as many people as infectious diseases. Immunizations and antibiotics are often touted as the greatest triumph of medical science, but the discovery of vitamins and the ensuing dietary improvements had a similar impact on public health.

To Europeans, scurvy was the deadliest of the vitamin deficiencies. It afflicted millions and determined the outcome of military battles and the fate of empires. It confounded physicians and bureaucrats. Despite the explorer Vasco da Gama and his men recognizing that eating fresh oranges cured the crew of scurvy in 1498, it took more than three hundred years for the Royal Navy to mandate the provision of citrus juice to its sailors. It took another hundred years to understand that scurvy is a nutritional deficiency disease.

It is astounding that it took centuries to grasp what now appears obvious. Since ancient times, people knew that they needed enough food to provide them with energy and that general malnutrition could kill. People also recognized the need for protein, thought to be necessary to repair tissues. However, people failed to understand the requirements for specific nutrients, some in only minuscule amounts. Physicians of that era, like today, were limited by the modes of thought at the time. The path from Vasco da Gama to the knowledge of the cause of scurvy provides many examples of prevailing wisdom leading people astray.

HOMING IN ON THE PRIZE

Once scurvy was proven to result from a faulty diet, the race was on to find the missing nutrient. The second part of the history of vitamin C describes the discovery of vitamins, the chemical characterization of vitamin C as the ascorbic acid molecule, and the large body of work defining its role in physiology. Progress was rapid once scientific medicine replaced speculative theories and the dogma of ancient philosophers. During the first decades of the twentieth century, nutritional science emerged as a major focus of medical research. It attracted the public eye and ambitious investigators. Academic prestige and Nobel Prizes were at stake, and big egos rose to the challenge. Brilliant science marked the story, as did pure luck, baffling wrong turns, and misunderstandings.

MEGAVITAMINS: THE CURE-ALL

In the 1970s Linus Pauling took up the cause of vitamin C. Pauling, a brilliant chemist and a charismatic personality, won the 1954 Nobel Prize in chemistry for adapting quantum physics to explain the chemical bonds that hold molecules together. During the 1950s and 1960s, he became a celebrity, touring the country to publicize the dangers of radioactive fallout from nuclear bomb testing. His efforts led to the treaty banning atmospheric nuclear testing and to winning a second Nobel Price, the Peace Prize, in 1962.

He then turned to another public health issue. He formulated a speculative theory that he called "orthomolecular medicine." He postulated that diet and vitamins could cure almost any disease and claimed that megadoses of vitamin C could prevent colds, influenza, cancer, and heart disease. Not content to publish his ideas in specialized scientific journals, he mounted a public crusade. Popular magazines and tabloid newspapers disseminated his ideas, and his book *Vitamin C and the Common Cold* was a best seller.

Although his claims were not backed by science, his evangelism bore fruit. Vitamin C sales skyrocketed and remained robust. The success of vitamin C propelled the expansion of the vitamin and supplement industry. How the industry evaded government regulation and grew to $40 billion in sales is a lesson about the relationship between science and politics and an example of how science is frequently widely ignored. It is a tale directly relevant to the twenty-first century, as climate change and a coronavirus pandemic again put politics and science in conflict.

WHY IT MATTERS

If the human intellect functioned perfectly, scientific knowledge would progress smoothly. Intelligent, trained investigators would build on the findings of their predecessors and seamlessly contribute to a growing edifice of understanding. The reality is not so neat. Scientists have egos, ambitions, rivalries, bosses, families, and financial interests, all of which can get in the way of the search for truth.

Scientists have preconceived notions about how the world works. These preconceptions range from mere hunches to established theoretical models. When correct, they provide a direct path to the right answer. But when wrong, they lead to the misinterpretation of data and fruitless experiments. These preconceived notions hamstring our thinking. As the story of vitamin C shows, they prevent smart people from seeing the obvious.

Scientists also work within the culture of their time and place, with its assumptions, prejudices, means of communication, and ways of thinking. That culture molds the way scientists think about the world. In the eighteenth century, a classical education taught men that the ancient sages had reached the pinnacle of wisdom.

Few thinkers of the day could break out of the confines of that teaching. Today's culture has its own confines. Thought leaders shape our view of the world, and public policies determine the research projects that receive financial support.

Food also carries strong emotional baggage. Food can take on almost magical qualities, with ritual and symbolic significance in religions. Ethnic cuisines are part of cultural identity. Food fads are prevalent. At some level, we hold on to the primitive idea that we are what we eat. Hence, nutrition has an emotional impact far beyond that of other medical topics. That emotion colors our reactions to the science behind it.

The saga of vitamin C illustrates these barriers to understanding, as well as how, with persistent effort, they can be overcome. This book traces a scientific enterprise across five centuries. It illustrates both triumphs and failures, from the first recognition of scurvy among sailors to ongoing molecular investigations.

I

BUCCANEERS AND BUREAUCRATS: THE HISTORY OF SCURVY

A DISEASE
OF MARINERS

Want to learn to pray? Then go to sea.

—sixteenth-century Spanish proverb[1]

THE LONGEST VOYAGE AND A NEW DISEASE

On a hot July day in 1497, four ships carrying about 170 men under the command of Vasco da Gama set sail from Lisbon bound for India.[2] King Dom Manuel was not willing to leave his cool mountain retreat to see the fleet off, but nobles wished da Gama and his men Godspeed from the dock. The citizens of Lisbon lined the banks of the Tagus River to cheer the ships on their way. They hoped the expedition would make Portugal a commercial power by breaking the monopoly that Genoa and Venice held as the European ends of the Silk Road.

Da Gama's expedition culminated efforts that had begun during the reign of Henry the Navigator (1394–1460). With the support of the Crown, Portuguese ships inched their way down the west coast of Africa. In 1488, Bartolomeu Dias and his crew became the first Europeans to round the Cape of Good Hope, but his crew, afraid to venture farther into unknown waters, forced him to return home without exploring beyond the tip of the continent. Nevertheless, by locating the bottom of Africa, he instilled confidence that the Indian Ocean was not enclosed by land, as fanciful Greek maps had depicted. He confirmed that a sea route from the Atlantic to Asia existed.

In 1483, King João II and his advisers met with Columbus, who proposed a path to the East Indies by sailing west. The Portuguese were unimpressed with Columbus and knew that he underestimated the circumference of the Earth. Having knowledge of the coast of Africa, they placed their bets on sailing east. So King João II and his successor, King Dom Manuel, charged Vasco da Gama with forging a route around the southern tip of Africa to India.

We know little of the early life of Vasco da Gama nor why the kings chose him rather than Dias to lead the expedition. Da Gama was only in his thirties, but he had seafaring experience and was a member of minor nobility with unquestioned loyalty to the Crown. Arrogant and quick to anger, he was a forceful leader with the ambition and courage to venture into the unknown.

An anonymous member of da Gama's crew wrote an account of the expedition.[3] "They left on their errand with the blessings of the Church, in the favor of their king, and amidst the acclamations of a sympathizing people." They sailed down the coast of Africa and landed at the Cape Verde Islands. One ship, carrying Dias, returned to Lisbon. The three others, two carracks and a smaller supply ship, set off for Asia. As the journalist states laconically, "On Thursday, August 3, we left in an easterly direction."

What he did not say was that after the ships crossed the equator, the commanders made an audacious decision. Rather than continue southward, hugging the coast of Africa and remaining within known waters, they veered away from land into the open Atlantic Ocean, probably coming within six hundred miles of the coast of

Brazil. They took this circuitous route to capture the prevailing westerly trade winds in hopes of being propelled around the Cape of Good Hope.

Why did they think this would work? As far as we know, Bartolomeu Dias and his men were the only Europeans to have sailed to the far southern Atlantic coast of Africa in their 1488 voyage. Dias had veered somewhat to the west and was able to take advantage of the trade winds to carry him around the cape. Presumably, da Gama sailed almost to South America before encountering westerly winds. It was an act of remarkable daring.

His daring was not rewarded. His ships did not round the tip of Africa on the first attempt but struck land just north of the cape. They had been out of sight of land for ninety-three days, likely the longest uninterrupted ocean voyage ever undertaken at that time. The men were suffering from dysentery, fever, and a shortage of drinking water. They collected water, meat, and fish upon landing; however, no fruits or vegetables were to be found. Fighting unfavorable winds and weather, they eventually rounded the Cape of Good Hope and landed on the east coast of Africa.

They then pushed on beyond the point at which Dias's crew had been afraid to venture. This demanded incredible courage. The first astronauts were unquestionably brave, but they knew what they were facing. Vasco da Gama and his men knew nothing of the Indian Ocean or of the inhabitants of the east coast of Africa. They had no idea if the locals would greet them warmly as visitors from exotic lands or attack them as hostile invaders. The men's willingness to venture beyond their known world reflects da Gama's qualities as a leader, as well as their hopes of sharing in the riches they expected to bring back from Asia.

As they sailed north, a mysterious illness afflicted the crew. Da Gama's journal provided the first clear description of this illness that was to plague mariners for centuries. "Many of our men fell ill here, their feet and hands swelling, and their gums growing over their teeth, so that they could not eat." The unknown disease ravaged the crew and men began to die. The explorers had no experience with such an ailment, no name for it, and no idea of what caused it. The ships anchored in the delta of the Zambezi River in Mozambique on January 22, 1498, almost six months after setting sail from Cape Verde.

In Mozambique, there were abundant fruits, especially oranges, growing along the river. The men ate the fruit and quickly recovered. "It pleased God in his mercy that on arriving at this city all our sick recovered their health, for the air of this place is very good." In keeping with the beliefs of the time, the narrator blamed the disease on the foul air aboard the ship. But the crew immediately recognized the value of oranges and eagerly ate them.

The three vessels proceeded along the African coast. Da Gama's arrogance in dealing with the local leaders led to several confrontations, some violent, making the voyage more adventurous than it had to be. They eventually reached India, where trading with the local merchants was again marked by conflicts and misunderstandings but no violence.

On the return voyage, they ignored warnings about monsoon winds and struck out on a direct route across the Indian Ocean. And once again the unknown illness reappeared.

> Owing to frequent calms and foul winds it took us three months less three days to cross this gulf, and all our people again suffered from their gums, which grew over their teeth, so that they could not eat. Their legs also swelled, and other parts of the body, and these swellings spread until the sufferer died, without exhibiting symptoms of any other disease. Thirty of our men died in this manner—an equal number having died previously—and those able to navigate each ship were only seven or eight, and even these were not as well as they ought to have been. I assure you that if this state of affairs had continued for another fortnight, there would have been no men at

all to navigate the ships. We had come to such a pass that all bonds of discipline had gone.[4]

On January 7, 1499, they anchored at Malindi, now part of Kenya, and the locals sent emissaries aboard. "The captain-major sent a man on shore with these messengers with instructions to bring off a supply of oranges, which were much desired by our sick. These he brought on the following day, as also other kinds of fruit." The men eagerly ate the oranges. Some were too ill to benefit from the fruit and died; the others recovered rapidly.

Too few men had survived to sail all three ships, so they burned the supply ship and set sail for home in the two carracks. When they rounded the cape on March 20, the remaining crew was in good health.

Da Gama arrived in Lisbon at the end of August 1499. He had lost two-thirds of his crew, mainly to scurvy. His brother, Paulo, who commanded one of the carracks, died on the return voyage. Da Gama had antagonized many of the rulers of Africa and India with his high-handed ways, but in Portugal he was a hero. Exultant men and women of Lisbon again lined the streets to cheer as princes and clergy escorted da Gama to the palace and an audience with King Manuel. The monarch heaped honors and riches upon da Gama but could not grant da Gama's wish to be given his hometown of Sines as his personal property; it was already the property of a powerful duke. Despite this disappointment, da Gama spent the rest of his life living like a king. And Portugal exploited its newfound access to Asia to dominate that trade. By the end of the sixteenth century, tiny Portugal was second in wealth only to Spain among the European nations. Although Columbus had been wrong about the circumference of the earth, the treasure that Spain looted from the New World was even more valuable than the Asian trade in silks and spices.

AN ERA OF TRADE AND COLONIZATION

Da Gama's voyage, along with that of Christopher Columbus five years earlier, opened the oceans to European nations and ushered in the Age of Sail. For the next 350 years, thousands of ships crossed the seas at the mercy of the winds. The vessels connected Europe to the rest of the world and enabled the construction of global empires.

The voyage of Vasco da Gama provided the first record of ships sailing on the open ocean continuously for more than a few weeks. Columbus on his first voyage to the West Indies required only thirty-seven days to cross the Atlantic. Before Columbus, sailors had remained close to land and touched shore frequently. European ships had been too small to carry enough men and provisions for months at sea and were not sufficiently seaworthy to survive ocean storms. China had built huge ships, more than five times larger than the biggest European vessels, but they made relatively short hops from port to port around the Indian Ocean. Those ships were designed to flaunt wealth and power, not to sail the open seas. Since the Chinese believed that they were at the center of the universe, they had little interest in exploring beyond familiar waters.

Another limitation to ocean travel was the primitive means of navigation, essentially limited to looking up at the sun during the day and the stars at night when the clouds permitted. Maps were mere sketches at best and pure fantasy at worst. To maintain their bearings, mariners required landmarks. They had to come within sight of land to orient themselves to known geography.

As a result, Europe had been a virtual island. The ocean was to the west, ice closed off the north, and Muslim realms and hostile warrior tribes blocked the south and east. The Crusades had penetrated North Africa, but the Sahara posed a barrier to travel farther south. Since Marco Polo in the thirteenth century, few Europeans had ventured into the Asian plains. Ambitious monarchs yearned to break through these barriers.

The impetus was greed. A thriving commerce brought Asian silks and spices to Europe over the Silk Road, a trade monopolized by Venice and Genoa. Other European nations wanted to muscle in on this lucrative business. They coveted direct access to Asian markets and luxury products without having to pass the goods through a series of middlemen, each marking up prices and extracting a share of the profits. That direct access required long ocean voyages.

Technological advancements fueled these ambitions. Beginning with the Vikings, ship construction had steadily advanced. Progress accelerated with the development of internal frames, which permitted the construction of large ships. The Portuguese had developed the caravel, a seaworthy craft slightly less than sixty feet long and, with fore-and-aft triangular sails, able to sail close to the wind. The Portuguese used the caravel to explore the west coast of Africa. Da Gama's two primary ships were carracks, larger and sturdier than the caravels, but fitted with traditional square-rigged sails, sacrificing maneuverability for size and seaworthiness. Because of this, da Gama had to sail almost to South America before he could capture the trade winds and turn east.

While ships became larger and more seaworthy, the development of a reliable ship's compass revolutionized navigation. The Portuguese also benefited from Spain's expulsion of Jews in 1492. The Jewish astronomer Abraham Zacuto fled to Portugal and brought with him detailed maps of the stars. With the newly invented mariners' astrolabe, a device to measure the angle of the sun and stars above the horizon on the rolling deck of a ship, the charts enabled sailors to calculate their latitude. These innovations emboldened Europeans to set out on the open sea. The two Iberian nations were the first to venture beyond the virtual European island, and Portugal was the more adventurous.

A DISEASE OF SAILORS

Although Vasco da Gama demonstrated that long sailing voyages over the high seas were feasible and gave Europe access to the world, the length of the voyage required the crew to live on ship's rations for months at a time. We do not know exactly what da Gama's men ate while at sea. Without refrigeration, their menu certainly lacked fresh fruits and vegetables.

The journal of da Gama's voyage provides the first detailed description among seamen of an illness characterized by lassitude, swollen gums, foul breath, swollen and painful extremities, and eventual death. Although the journal blamed the disease on bad air, the men recognized that oranges were a cure and ate them at every opportunity. Having no concept of vitamins, they did not realize that the fruit replaced a vital nutrient, vitamin C, of which they had been deprived for months.

This malady soon became known as *scurvy*, a word derived from the Icelandic word *skybjugr*, meaning cut or ulcerated swellings.* An alternative English spelling was *scorby*, and people suffering from scurvy are *scorbutic*. In Portuguese and Spanish, the disease is *escorbuto*. Since vitamin C is the antiscorbutic vitamin, its chemical name became *ascorbic acid* or *ascorbate.*†

Although da Gama and his men had never seen the disease before, there had been reports of scurvy before the Age of Sail. Ancient Greek and Egyptian physicians described soldiers and civilians suffering from malnutrition during sieges. They had symptoms of scurvy mixed with features of other diseases. In northern Europe during the Middle Ages, scurvy afflicted the population each win-

---

*The word *skybjugr* entered Danish as *scorbuck* and was Latinized to *scorbutus*. In sixteenth-century English writings it became *scorby* or *scarby* and then *scurvy*. This is a different etymology than the pejorative use of the word, as in "he's a scurvy dog."

†Vitamin C may be either of two compounds, ascorbic acid or dehydroascorbic acid. These two compounds differ by two hydrogen atoms and can be interconverted within cells. For the purposes of this book, vitamin C and ascorbic acid are used synonymously.

ter when fresh vegetables were unavailable and before the potato, which is rich in vitamin C, arrived from the New World. The Crusaders suffered from scurvy and other forms of malnutrition when they invaded North Africa without adequate supply lines.

However, scurvy was not recognized as a specific disease and a prominent concern of medical writers until it appeared among mariners.[5] It was perhaps the first disease of technology, a product of innovation in ship construction and navigation, which allowed ships to remain at sea for months at a time.

After about three months of a diet lacking vitamin C, the first signs of scurvy appear. The initial manifestation is lassitude, which becomes profound as the disease progresses. Next, tiny bumps appear around hair follicles, and the hairs develop an abnormal corkscrew shape. The disease is painful; the joints, muscles, and back ache. Tiny hemorrhages, known as petechiae, appear in the skin when capillaries rupture.

As the disease progresses, connective tissues break down. New wounds fail to heal and old ones may open up. Hemorrhages cause swelling of the arms and legs. Blood can leak into the spaces between the bones and the tough connective tissue lining the bones, the periosteum; this type of hemorrhage is virtually unique to scurvy, excruciatingly painful, and a prominent feature of scurvy in infants. Bleeding into muscles and other organs can also occur.

Among people with poor dental hygiene, the gums become soft and swollen, and the teeth loosen in their sockets and eventually fall out. The breath smells horrible. Physicians thought that the gum changes were required to make the diagnosis. It wasn't until the late nineteenth century that doctors realized that people who have lost their teeth, infants who have not yet teethed, and adults with good dental hygiene manifest few or no gum changes.

The combination of profound lassitude, pains throughout the body, and the inability to eat solid food was agonizing. The disease can be fatal within six months, although the exact mechanism by which death occurs is unknown. In some, hemorrhages occur in the heart muscle or in the lining of the heart. Those with advanced disease may faint if they try to stand and may collapse and die suddenly. At that point, death is a merciful end.

THE PLIGHT OF THE COMMON SAILOR

The technological advances that ushered in the Age of Sail also ushered in the age of colonial empires. European nations exploited the ability to travel long distances not only to trade with distant nations but also to acquire colonies to serve as sources of slaves and raw materials and to provide captive markets for their manufactured goods. England, Spain, Portugal, and the Netherlands were the major colonial powers of this era. They built large merchant fleets and maintained navies to protect them. These fleets demanded the services of hundreds of thousands of sailors.*

The lives of these sailors were unimaginably brutal. The men were imprisoned onboard and not permitted to go ashore while the ship was in port for fear that they would desert. The sailors performed arduous physical labor in four-hour shifts (watches) around the clock. When not on duty, they lived in crowded quarters infested with vermin and insects. The ships had poor ventilation, and everything, including the clothing and bedding, was constantly wet. The stench was terrible. Ships stank from unwashed men crowded together for months at a time in unsanitary conditions, from rotting provisions, and from whatever grew in the bilge water sloshing below deck. In these conditions, infectious diseases, especially dysentery, were prevalent.

---

*This naval history relies heavily on British sources, as the English wrote more openly about their experience with scurvy than other nationalities. English sea captains liked to boast about their exploits. The British were also assiduous in maintaining records. Both the Royal Navy and the British East India Company kept detailed budgets and written accounts of their activities.

To allow the crewmen to perform their labors, they received a high-calorie diet rich in fat and protein. The typical rations of common seamen on an English naval vessel were:

Biscuit	half a pound, daily
Salt beef	two pounds, twice weekly
Salt pork	one pound, twice weekly
Dried fish	two ounces, three days weekly
Butter	two ounces, twice weekly
Cheese	four ounces, three days weekly
Peas	eight ounces, four days weekly
Beer	one gallon, daily

At times, they also received "portable soup," a dried cake made from oxen offal mixed with vegetables and preserved with salt.[6] When put in hot water, it provided an occasional warm dish. This diet was deficient in vitamin C, as well as vitamin A and the B vitamins thiamin and niacin. As a result, crews often suffered multiple vitamin deficiencies.

Within a century of Vasco da Gama, as long-distance sailing voyages became common, scurvy evolved from an unknown disease to a common affliction of sailors. Sir Richard Hawkins, who led an expedition from England around South America from 1592 to 1593, wrote an especially vivid account of its devastation.[7] Hawkins claimed that in his twenty years at sea, he had personally seen ten thousand sailors with the disease. His comment highlights how common scurvy had become and the extent to which commanding officers took it for granted.

Hawkins was from a family with a strong naval tradition. He was the son of an admiral and had extensive seafaring experience, including commanding a ship that fought the Spanish Armada in 1588. The cover story for the public was that his expedition to South America was to discover and map new territories, but the true intent was piracy. He set out to disrupt the commerce of Spain by capturing its merchant ships and plundering its settlements along the west coast of South America.

Three ships and about two thousand men set sail June 12, 1593. Facing unfavorable winds, they did not land in Brazil until October, by which time scurvy afflicted the crew and only twenty-four men on the flagship were healthy enough to man the sails. Hawkins described the primary features of the disease, "a loathsome slothfulness . . . swelling of all parts of the body, especially of the legs and gums, and many times the teeth fall out of the jawes without paine." He attributed the disease to multiple causes, including bad air aboard ships, spoilage of the ship's rations, and a weak stomach with poor digestion brought on by a hot climate.

Hawkins believed that scurvy could be prevented by keeping the ship clean; wholesome exercise; eating as little salted meat as possible when in a hot climate; not curing meats or washing garments in salt water; and wearing clean, dry clothing, especially when sleeping. He failed to mention how to accomplish this on ships crammed with two thousand men. He also stressed the importance of diet and advised that every morning each man on board should be given a bit of bread and a drink of either beer or wine mixed with water. But "That which I have seen most fruitfull for this sickness is sower [sour] oranges and lemmons." He claimed he had less scurvy on his ship than on the other two during the expedition because he gave the men lemon juice until the supply ran out.

Following the beliefs of the times, he attributed scurvy to bad air. "But the principall of all, is the ayre of land; for the sea is natural for fishes, and the land for men. And the oftener a man can have his people to land, not hindering his voyage, the better it is, and the profitablest course that he can take to refresh them." He followed this practice on his voyage and touched shore often. Although he thought the good air to be the main benefit of periodic landings, he also took every opportunity to collect fresh fruits.

He purchased between two and three hundred oranges and lemons in Santos, Brazil.

> Comming aboord of our shippes, there was great joy amongst my company; and many, with the sight of the oranges and lemmons, seemed to recover heart. This is a wonderfull secret of the power and wisdome of God, that he hath hidden so great and unknown vertue in this fruit, to be a certaine remedie for this infirmitie; I presently caused them all to be reparted [shared] amongst our sicke men, which were so many, that there came not above three or foure to a share.[8]

This echoed the experience of Vasco da Gama, whose crew eagerly ate oranges whenever they were available. Like da Gama, Hawkins recognized that oranges and lemons were curative. That the fruit provided an essential nutrient did not occur to him, and bad air remained the culprit.

Despite his efforts, among Hawkins's three ships, which had set out with about two thousand men, only seventy-five survived when they reached Valparaiso, Chile, in April 1594. Almost all the losses were from scurvy. Even with his depleted manpower, he captured several Spanish ships and looted two towns. Finally, off the coast of Ecuador, the Spanish captured Hawkins and his few surviving men. Hawkins spent several years in a Spanish prison but eventually found his way back to England. In 1622 he published his account of his adventures.

At the time, merchants like Hawkins were more assiduous than the Royal Navy in providing citrus juice to their crews. In 1600, the British East India Company sent three ships commanded by Sir James Lancaster to trade with Sumatra. Ten years earlier, Lancaster had led England's first commercial voyage around the Cape of Good Hope into the Indian Ocean. It was a disaster. He started with four ships and returned with only one manned by a crew of five men and a boy.[9] Among other horrors, scurvy devastated his crew.

An anonymous author, presumably an officer aboard Lancaster's ship, wrote an account of Lancaster's second voyage.[10] The first leg of the trip resulted in an unintended experiment. When the three ships landed at the Cape of Good Hope, Lancaster's ship had less scurvy than the other two.

> And the reason why the general's [Lancaster's] men stood better in health than the men of other ships was this; he brought to sea with him certaine bottles of the juice of limons, which he gave to each one as long as it would last, three spoonfuls every morning fasting, not suffering them to eat anything after it till noone. This juice worketh much the better if the partie keepe a short dyet, and wholly refraine salt meat; which salt meat and long being at the sea, is the only cause of the breeding of this disease.[11]

When they landed along the coast of Africa, the officers made every effort to obtain oranges, knowing that they would cure the men of scurvy. Like Hawkins and da Gama before him, they knew how to prevent and cure the disease, but they remained ignorant of the reason.

During the early 1600s, English and Dutch merchant ships carried lemon juice for the sailors. For this purpose, the Dutch East India Company maintained lemon plantations in Mauritius and South Africa. In 1617 John Woodall, the first surgeon general of the English East India Company, published *The Surgeon's Mate*, a textbook of nautical medicine. He advised that oranges, lemons, and limes be taken on all prolonged sea voyages. "The use of the juyce of Lemons is a precious medicine and well tried being sound and good."[12] Usually, the juice was saved to treat cases of established scurvy and used as a preventative only if stores were abundant.

Even as a preventative, it was not uniformly effective. The juice lost its potency after a few weeks' time in storage.

Woodall's advice was ignored. By the middle of the 1600s, partially due to its expense and partially due to its inconsistent benefits after storage, the provision of lemon juice was rare, and scurvy remained the most common cause of death of seamen. In the late seventeenth and early eighteenth century, it mainly affected merchant ships, as naval vessels often remained close to shore, protecting their home countries. Later in the eighteenth century, a shift in tactics led to naval ships remaining at sea for prolonged periods, blockading enemy ports or patrolling hostile shores where they could not land and reprovision. This made scurvy the scourge of both naval and merchant seamen.

The loss of men was astonishing. During the Seven Years' War from 1756 to 1763, of the 184,899 men who served in the Royal Navy, 133,708 were listed as lost, either to disease or desertion. A detailed breakdown of the diseases was not provided, but scurvy was the most frequent. During the American War of Independence from 1774 to 1783, a total of 170,910 men served in the Royal Navy's West Indies Fleet, which patrolled the Caribbean and South Atlantic. Of these, 1,243 died in action, whereas 18,545 died of disease. In the tropics, most of these deaths were a result of infections, but four to six thousand were due to scurvy. However, the biggest loss was from desertion: almost a quarter of the sailors—42,069 men—took the opportunity to escape the miserable conditions aboard the ships.[13]

Scurvy could determine the outcome of sea battles. It is likely that the defeat of the Spanish Armada was at least in part a result of the debilitation of the Spanish sailors by scurvy, having been confined to their ships eating a sailor's diet longer than the British. Although the Spanish did not specifically document the ravages of scurvy, they did note that when the armada returned to Spain, more than four thousand sailors were on the sick rolls, most suffering various forms of malnutrition.[14]

Scurvy also occurred on land, especially in prisons, and in both military and civilian populations during prolonged sieges. Although it occurred in northern Europe during the winter, it did not afflict the indigenous people of North America. They knew how to stave off the disease. The Inuit ate raw seal liver, which contains large amounts of vitamin C, and other indigenous peoples drank infusions prepared from evergreen trees, a source of the vitamin, during the long winters.

This knowledge saved one group of European explorers. In 1535, Jacques Cartier led a French expedition up the Saint Lawrence River searching for the Northwest Passage. When forced to overwinter in northern Quebec, scurvy emerged among his men. Out of 110 men, all but three suffered from the disease and twenty-five died. From a chance encounter, Cartier learned from a native that a tea prepared from an evergreen tree, perhaps an arborvitae, would cure the disease. Taking advantage of this wisdom, the remaining members of the expedition survived the winter. Apparently, Cartier did not pass this knowledge on to subsequent French expeditions to Canada, as they continued to suffer from the disease.[15]

THE CONTINUING MENACE OF SCURVY

It is estimated that prior to 1850 the Royal Navy lost more than one million men to scurvy. Other European nations did not catalog their losses so carefully, but they also suffered tremendous attrition from the disease. In view of the belief of many commanders that citrus fruits could cure scurvy, it seems incredible that authorities did not institute effective preventive measures. There were several reasons for this failure.

First and foremost, no one understood the mechanism by which scurvy afflicted sailors. Before the late nineteenth century, general malnutrition was understood, but there simply was no concept of a disease being caused by a deficiency of one specific nutrient. The idea that a person could suffer a nutritional deficiency while being supplied with plenty of food and without losing weight was alien.

When the early observers saw scurvy being cured by lemons and oranges, they looked for explanations that fit with theories of disease prevalent at the time. Diseases were all thought to result from an external agent acting on the individual. That eating fresh citrus fruits cured scurvy did not prove that the disease was nutritional. We know antibiotics cure infections, but we also know that infections are caused by microorganisms, not by a nutritional deficiency of antibiotics.

The prevailing theory of disease was the humoral theory, which harkened back to second-century Greek writers Hippocrates and Galen. The theory posited that there were four humors: blood, phlegm, black bile, and yellow bile. When an individual was healthy, these four humors were pure and in balance. Disease resulted when something external to the individual contaminated or upset the balance of these humors. The humoral theory was prevalent in European medical teaching until the eighteenth century.

One idea of the time was that bad air, or miasma, could poison the humors. Miasma emanating from damp soil was thought to explain why Europeans suffered so many diseases when traveling to the tropics. Human-to-human contagion was understood in principle, but transmission of microorganisms by insects or other environmental vectors was not understood until the discoveries of Pasteur, Koch, and Lister in the late nineteenth century. Hence, the tropical air was made the culprit. The concept that bad air caused disease was popular from the sixteenth century on and received a boost in the late eighteenth century when the discovery of oxygen and other atmospheric gases piqued interest.

The conditions aboard sailing ships were indeed abysmal, and the air was especially bad. Knowing nothing of microorganisms or vitamins but directly experiencing the stench, writers blamed the foul-smelling air for causing disease. The oranges were assumed to counteract the effects of the air in some unknown manner.

A consistent observation of commentators was that scurvy was a disease of common seamen; officers rarely suffered from the illness. Given the strong class system of the times, the common seamen—all from the lower classes—were blamed for their own misery. Writers invoked traits such as laziness, poor hygiene, and debauchery to common sailors, rendering them susceptible to diseases the upper classes escaped. The officers never thought their more varied diet and freedom to go ashore when in port might be the explanation.

Finally, prior to the invention of refrigeration, there was no way of storing fresh fruits and vegetables on ships during long voyages, and vitamin C in citrus juice lost its efficacy with storage. Merchant ships were at the mercy of the winds, and their captains were reluctant to waste time by detouring to ports to take on fresh supplies. Naval vessels were frequently patrolling hostile shores for months at a time, and it would have been expensive and risky to shuttle provisions to them using supply ships.

Because of these misunderstandings, mariners and those on land continued to suffer and die from a deficiency of vitamin C. Nothing could be done without a new approach to understanding the disease.

2

CATASTROPHE AND ENLIGHTENMENT

*Of theory in physic the same may perhaps be said,
as has been observed by some of zeal in religion.
That it is indeed absolutely necessary; yet, by carrying it too far, it may be doubted whether it has
done more good or hurt in the world.*

—James Lind, *A Treatise on the Scurvy*, 1753

A NAVAL DISASTER

Commodore George Anson led a British expedition that circum-navigated the globe from 1740 to 1743 and inflicted economic damage on Spain. The press hailed it as a great victory. However, scurvy ravaged the fleet, killing more than three-quarters of the crew. This devastation brought the disease to the attention of the public and led to a crucial step in its understanding the illness.[1]

Anson was born in 1697 to an aristocratic family in Staffordshire, England. He volunteered for the navy at age fifteen and rapidly rose through the ranks, commanding his first ship at age twenty-five. At age forty he achieved the rank of commodore. When England declared war on Spain in January 1740, the Admiralty charged Commodore Anson with organizing an expedition to the Pacific to capture the Spanish treasure galleon that shuttled silver and gold from Acapulco to the Philippines and returned with Asian silks and spices for transshipment to Europe. To avoid the dangerous sea voyage around Cape Horn, the Spanish carried these precious goods over land from Acapulco to Veracruz for shipment on to Spain.

Anson set sail on September 18, 1740, with a fleet of six warships and two supply ships, together carrying almost 1,400 crew members and 500 marines.[2] The flagship was the *Centurion*. An account of the expedition was later published under Anson's name, although it is not known who actually wrote it. To treat scurvy, Anson took a supply of Ward's pills, a patent medicine containing antimony and balsam, poisons that induced sweating and vomiting. Their inventor, Joshua Ward, touted them as a panacea that would rid the body of noxious substances and cure any disease, including scurvy, syphilis, and cancer.

By the time the fleet reached Cape Horn in March 1741, scurvy had afflicted the crew. Men soon began to die, the death rate reaching six to eight per day. The depleted and debilitated crews fought to make the stormy passage around Cape Horn, but three warships had to turn back. One supply ship had already turned back from Brazil after offloading its cargo to the other ships.

On June 9 the *Centurion* arrived at Juan Fernandez, the Pacific island on which Alexander Selkirk, the presumed model for Robinson Crusoe, had been marooned. Two warships and the remaining supply ship that had managed to round the cape later joined the *Centurion*. The badly damaged supply ship was broken up and its crew transferred to another ship. The three remaining warships had departed from England with a total of 961 crew, of which 335 remained alive at the beginning of September. By far the greatest mortality had been from scurvy. By December 7, 1741, only 201 survived.

Despite the depleted manpower, they captured several Spanish merchant ships and looted a town on the coast of Chile. They left the coast of South America May 6, 1742, bound for China. After seven weeks at sea, scurvy reappeared. Mr. Ward's pills did the job of inducing sweating, vomiting, and diarrhea but did not cure scurvy. The horrors did not abate; men continued to die. A single ship, the *Centurion*, was all that remained of the fleet, the other two ships having been irreparably damaged in battles and storms.

The *Centurion* at long last reached Formosa, where it captured a Manila galleon with its treasure of silver and gold. The survivors sailed on to England, arriving June 15, 1743, after circumnavigating the globe on a voyage lasting three years and nine months. Of the almost one thousand men who left England in the three main ships, only 188 made it back home. Almost all the deaths were due to scurvy. The *Centurion* arrived laden with the vast fortune from the Manila galleon, making Anson fabulously wealthy and his surviving crew all rich. Although poorly paid, the common seamen shared in the spoils of piracy.

The English, in their typical stiff-upper-lip fashion, feted the survivors as heroes for overcoming adversity and inflicting a great loss on the enemy. Not only was Anson not blamed for the loss of life, but he was promoted to First Lord of the Admiralty, head of the entire Royal Navy. However, the publicity surrounding the voyage and its harrowing death toll brought scurvy to the attention of the naval authorities and the public and motivated a modest ship's surgeon, James Lind, to publish the most famous work in the history of the disease.

THE NEW WAY OF THINKING

James Lind was born and educated in Edinburgh, which was the intellectual center of the Scottish Enlightenment during the middle decades of the eighteenth century.[3] Through the Middle Ages and Renaissance, European scholars held that human wisdom peaked with the classic Greek and Roman scholars and subsequently declined. The work of scholars was to explicate classical teaching and apply it to the questions of the day. Scottish thinkers such as David Hume and Adam Smith challenged this view, arguing that human reason and direct observation should replace reliance on ancient authorities.

David Hume published his influential work of 1748, *An Inquiry Concerning Human Understanding*, asserting, "In vain, therefore, should we pretend to determine any single event, or infer any cause or effect, without the assistance of observation and experience." This new way of thinking catalyzed the growth of science, the industrial revolution, and the economic rise of western Europe.

James Lind was born in 1716 to a middle-class mercantile family.[4] At the age of fifteen, he apprenticed with a local physician and attended lectures, without being a formal student, in the newly founded medical school of the University of Edinburgh. Although the medical teaching remained grounded in the four humors of the ancient Greeks, Lind came to believe that direct observation and experiment were essential to advancing knowledge.

In 1739, upon completing his apprenticeship, he entered the Royal Navy as a surgeon's mate. This was not a surprising career path. Although a surgeon's mate earned a meager salary and worked in miserable conditions on naval vessels, joining the navy allowed Lind to practice medicine without a university degree. Moreover, establishing a successful private practice would have required more society connections than Lind enjoyed.

England had just declared war on Spain, and Lind may well have been motivated by patriotism in addition to thoughts of furthering his career. Although Scottish nationalism smoldered throughout the first half of the eighteenth century, Lind probably considered himself more English than Scottish. He spent most of his professional life in England and, in 1794, was buried in an Anglican church in England, not a Presbyterian church in Edinburgh.

LIND'S EXPERIMENT

In 1746, Lind was promoted to surgeon and assigned to HMS *Salisbury*, a ship of fifty to sixty guns with a crew of about 350 men. The *Salisbury* belonged to the Channel Fleet, shielding Britain from invasion during the Napoleonic Wars. In April and May 1747, despite being at times in sight of land, scurvy appeared among the crew. On May 20, 1747, Lind chose twelve sailors afflicted with the illness and made them the subjects of his famous experiment.

He housed the men in the ship's sick bay where he could keep them under close observation. He gave them all an identical standard naval diet completely lacking vitamin C: a breakfast of gruel sweetened with sugar, a lunch of either fresh mutton broth or puddings and boiled biscuits with sugar; and a supper of barley and raisins, rice and currants, sago, and wine. He grouped them into pairs and gave each pair one of six supplements:

- One quart of cider daily
- Twenty-five drops of vitriol (sulfuric acid) three times daily
- Two spoonfuls of vinegar three times daily
- A half pint of seawater daily
- Two oranges and one lemon daily for six days (when the supply ran out)
- A nutmeg-sized paste of garlic, mustard seed, dried radish root, balsam of Peru, and gum myrrh three times daily

Lind reported the unambiguous results:

The most sudden and visible effects were perceived from the use of the oranges and lemons; one of those who had taken them, being at the end of six days fit for duty. The spots were not indeed at that time quite off his body, nor his gums sound; but without any other medicine, than a gargarism of elixir vitriol, he became quite healthy before we came into Plymouth, which was on the 16th of June. The other was the best recovered of any in his condition; and being now deemed pretty well, was appointed to nurse the rest of the sick.[5]

The only other pair that seemed to benefit were those who drank the cider. When the experiment was discontinued after a fortnight, they had some improvement in their gums and lassitude. The elixir of vitriol helped the gums and mouth but not the other symptoms.

This experiment became famous because it was one of the first controlled clinical trials in the history of medicine. Lind directly compared groups of experimental subjects treated identically except for one treatment variable, in this case the dietary supplement. He followed many of the rules that survive to this day for such experiments. Lind chose men all suffering from the same disease and all with a comparable degree of severity. He divided them into groups and treated them all identically. He housed them in the same room on the ship, sharing the same air and germs. He kept them under close observation so that the men could not wander around the ship or get food from their friends. He gave them an identical diet. The only thing that differed was the supplement to that diet. If there was a difference in outcome among the groups, it could be due only to that one difference.

Even though there were only two sailors in each group and a single experiment is rarely sufficient to prove a hypothesis, the result appeared inescapable. Oranges and lemons cured scurvy, and the other treatments did not. The experiment was a milestone in the history of medical science.

THE TREATISE

Lind resigned from the navy in 1748 and entered private practice in Edinburgh. While engaged in his practice, he wrote the first edition of his *Treatise on Scurvy*, which he published in 1753. Perhaps with an eye to advancing his career, he dedicated the book to George Anson, who by then was First Lord of the Admiralty.

The full title was *A Treatise on the Scurvy in Three Parts. Containing An Inquiry into the Nature, Causes and Cure, of That Disease, Together with a Critical and Chronological View of What Has Been Published on the Subject.* As the title promised, Lind began with a scholarly and critical review of past writings. He summarized previous descriptions of scurvy going back to Hippocrates and Pliny. Lind does not mention Vasco da Gama or Richard Hawkins and thought that the first credible description of the disease was that of Jacques Cartier, who, as previously mentioned, led a group of colonists to Hudson Bay in 1535 and suffered scurvy in the winter.

Lind reviewed the descriptions of other naval surgeons, who vividly described the misery of the disease. He discussed previous theories of the cause of scurvy but dismissed them all. He discounted the effect of the foul smell of the ships, salted food, and exposure to salt water. Despite the unequivocal results of his experiment, he also argued against the theory of Johann Friedrich Bachstrom, a Dutch physician who wrote in 1734 that "this evil is solely owing to a total abstinence from fresh vegetable food and greens; which alone is the true, primary cause of the disease."[6]

Lind did not completely discard the contribution of diet in causing scurvy but thought it was but one of several factors. He based his argument on the lack of scurvy in regions such as Scotland and Newfoundland, where the populace goes without fresh vegetables for half the year. Lind did not know that many populations in northern latitudes had alternate sources of vitamin C in the winter: potatoes in the British Isles, sauerkraut on the Continent, and raw seal liver and tea prepared from evergreens in the Arctic regions of North America.

Lind, like others before him, primarily blamed the air aboard ships, though Lind blamed its humidity not its foul odor. He stated, "the *principal and main predisposing cause* to it, is a manifest and obvious quality of the air, *viz.* Its moisture." The combination of cold and moisture "is the most powerful predisposing cause to this malady." Lind attributed other secondary causes to prior weakening by other maladies and the moral defects of common seamen, their lazy inactive disposition and discontented melancholic humor.

He also conceded that the want of fresh vegetables and greens was a secondary cause of the disease. "Experience indeed sufficiently shews, that as greens of fresh vegetables, with ripe fruits, are the best remedies for it, so they prove the most effective preventatives against it. And the difficulty of obtaining them at sea, together with a long continuance in the moist sea-air, are the true causes of its so general and fatal malignity upon that element." He could not conceive of a disease caused by a lack of a nutrient that did not provide either calories or protein. Consequently, he resorted to contorted reasoning to remain true to his teachings and the dominant theories of his times and explain how the cold, damp air could cause a disease that could be prevented or cured by eating fresh fruits and vegetables.

During his naval career, Lind saw hundreds of cases of scurvy while sailing the Mediterranean and the Channel. He drew on his own experience to give a graphic and gruesome depiction of the disease. He was wrong in thinking that the gum changes were essential in diagnosing the disease. People with good dental hygiene, without preexisting inflammation of the gums, or those without teeth do not develop the dental abnormalities. However, the men that Lind saw almost universally had the changes, and it was such a striking feature among sailors that it is understandable that Lind would have assumed it was a universal feature of the disease. Lind noted that the most severely affected sailors were frequently those who had been impressed into service, whereas officers were seldom affected by scurvy.

As was to prove as important to understanding the disease as his controlled experiment, Lind also described the postmortem findings of those who had died of scurvy. His descriptions were based on his own experience and that of others, including the surgeons with the Anson expedition.

He noted that frequently there were hemorrhages into muscles and that the blue, red, yellow, and black spots on the skin represented subcutaneous hemorrhages of different ages. He gave a gruesome description of bones coming apart and the ends rubbing together.

> In some, when moved, we heard a small grating of the bones. Upon opening these bodies, the epiphyses [the ends of long bones] were found entirely separated from the bones; which, by rubbing against each other, occasioned this noise. In some we perceived a small low noise when they breathed. In those the cartilages of the sternum were found separated from the bony part of the ribs.[7]

He noted that these bone findings were especially common in men younger than eighteen years of age. A hundred years later, these observations proved vital in recognizing scurvy in infants and still later in identifying the disease in animals.

To buttress the conclusion of his experiment, which was based on only two cases, he cited the experiences of several other voyages under different commanders during which oranges or limes cured the crew of scurvy. He cited instances of Dutch and English ships sailing together and the English ship being struck with scurvy while the Dutch sailors remained healthy. He recognized the benefit of sauerkraut in a footnote: "The Dutch sailors are less liable to the scurvy than the English, owing to this pickled vegetable (cabbage) carried to sea." He also noted that since French colonists in Hudson Bay started using spruce beer, which is rich in vitamin C, scurvy had been rare.

Lind made a major error in proposing a method of preserving the juice of oranges and lemons for long voyages. He called this a *rob*. He prepared the rob by placing a basin of juice into almost boiling water for several hours until it reduced to the consistency of a syrup. He advised that the rob be stored in bottles where it would keep for several years, meaning that it would retain its acidity and not show signs of microbial contamination. However, he never tested its antiscorbutic properties. We now know the prolonged heating would have led to oxidation and subsequent degradation of the ascorbic acid. In practice, the rob proved useless.

THE THEORY

Lind had attended lectures at the University of Edinburgh peopled with disciples of the Dutch physician Herman Boerhaave, who had tried to adapt the teachings of Galen and the ancient Greeks to the medicine of the day. Lind lamented that medicine relied more on theory than observation, but he followed in the footsteps of his teachers by producing a mashed-up theory of scurvy, conflating the ancient Greek theory of the balance of humors, miasma, and his own speculation about how citrus fruits might cure the disease.

He started with the belief that scurvy was a putrefactive disease. Putrefaction was the name given to the decomposition a body undergoes after death. If not preserved by embalming, a dead body begins to stink, it produces gases that cause it to swell, and the tissues turn to liquid. The theory held that the living body undergoes the same process, but it can repair tissues with the protein in food and excrete the noxious waste products in the urine, feces, or sweat. Since scurvy caused no obvious problem with urination or with the bowels, the sweat must be the culprit.

Lind postulated that cold, dampness, and lack of exercise predisposed sailors to scurvy by clogging their pores and preventing them from sweating out toxic substances. This blockage allowed the products of putrefaction to accumulate and poison the humors. To Lind, the rotting gums and foul breath of the scorbutic sailors were clear evidence of putrefaction.

To explain the efficacy of citrus fruits in curing the disease, Lind invented a complicated theory of digestion. He posited that digestion breaks down vegetables and fruits into substances that can be excreted easily in perspiration, thereby unclogging the pores. His formulation, although entirely wrong, was consistent with the accepted wisdom of the day.

Lind exhibited a universal feature of human thought. We will struggle mightily to shoehorn novel observations into our favorite theory before we will try inventing a new theory. Among scientists, this represents a conscious attempt to be conservative. They do not discard a theory that has served well until the evidence against it becomes irrefutable. But it is also a path of least resistance. It is easier to contort the evidence to fit our theories than to develop new ones.

THE UPSHOT

Lind's *Treatise* is widely lauded as a milestone, marking the beginning of a scientific approach to disease. Even though he could not completely break with past teaching, he was sufficiently imbued with the teachings of the Enlightenment to place his own direct observations and experiment above established authority. His clinical trial was a groundbreaking experiment that offered a cure for a fatal disease.* His descriptions of the clinical features and postmortem pathology of the disease proved essential in the eventual understanding of its mechanism. However, the *Treatise*, although widely read by physicians and published in two subsequent editions, had little practical impact at the time of its publication. The Admiralty made no change in its policies and ships' captains made no attempt to provide fresh fruits or vegetables for the crew. There were several reasons for that failure.

Some originate within the *Treatise* itself. Lind was a modest man, and his modesty led him to respect established authorities, even as he dismissed their theories. He spent many pages giving them their due. He went back to ancient texts to do a complete review of past thinking on the subject. His clinical trial occupies only a few pages of his text and is buried in the middle and ignored in the book's title. Then, after reporting his observations, rather than drawing the obvious conclusion that scurvy was a nutritional disease, he tried to fit his observations into the prevailing theories of disease that he had been taught at the University of Edinburgh. The bodily humors was the only theory of disease available to him. There was no existing paradigm of a deficiency of a specific nutrient into which Lind could fit his observations.

---

*Although often described as the very first controlled clinical trial, there were documented controlled trials in much earlier times, motivated primarily by commercial rather than scientific aims. See A. Rankin and J. Rivest, "Medicine, Monopoly, and the Premodern State—Early Clinical Trials," *New England Journal of Medicine* 375 (2016): 106–9.

Lind also misconstrued the evidence. He dismissed the dietary theory of scurvy based on the ability of large populations of people, including northern Europeans and the indigenous peoples of North America, to live up to nine months a year without fresh fruits and vegetables without developing scurvy. Lind did not understand that these people had foods that could replace fresh fruits and vegetables in their diets during winter months. This was to become tragically apparent in Britain during the Potato Famine of the next century.

In his own experience, sailors could remain free from scurvy during voyages of up to three years. He did not appreciate the value of shore leave, providing periodic access to fresh fruits and vegetables during long voyages. He acknowledged that sauerkraut, preserved for winter consumption by Dutch and Germanic populations, could be a replacement for fresh vegetables. We now know that sauerkraut retains just enough vitamin C to stave off scurvy. Also, many sailors brought raw onions with them, another good source of vitamin C.

Lind's most fateful error was his failure to test his rob. He assumed that if it were protected from bacterial growth and retained its acidity it would retain its antiscorbutic properties. In fact, heating citrus juice in the presence of air leads to oxidation and degradation of the ascorbic acid. When ships' surgeons failed to cure scurvy with the rob, they lost faith in the value of any citrus juice.

In addition, there were external factors that hindered the addition of citrus juice or fresh vegetables to the diet of sailors. One was bureaucratic inertia. The naval authorities fiercely resisted any added expense. The British Crown had limited ability to tax the population and raised most of its revenue from the poor through sales taxes. Almost all proceeds went to support the landed aristocracy and virtually continuous warfare. Sailors were paid a pittance and hoped to be compensated for their dangerous and grueling labor through their share of the treasure from captured enemy vessels, including merchant vessels. Much of what passed for naval warfare at the time was merely piracy.

Simple and obvious methods could have provided fresh provisions. Tenders could have reprovisioned the ships of the Channel Fleet, which frequently patrolled near land. However, that would have incurred added expense. Since scurvy was overwhelmingly a disease of ordinary seamen, there was little incentive for the Admiralty Board, also known as the Lords of the Admiralty and drawn from the British aristocracy, to devote money and effort to the well-being of men they looked down on as the dregs of society.

In 1758, likely through the good offices of George Anson (by then Lord Anson), Lind was appointed Physician to the Royal Hospital Hasler at Gosport, next to the huge naval base at Portsmouth. Hasler was the main naval hospital in Great Britain and soon to become the largest hospital in Europe. Despite the publication of his *Treatise*, cases of scurvy continued to flood into Hasler. During his first two years in charge, 1,146 of 4,275 admissions were for scurvy.

In an afterword to the 1772 third edition of his *Treatise*, Lind detailed his treatment regimen for these cases: four-and-a-half ounces of lemon juice and two ounces of sugar in a pint of wine consumed every twenty-four hours. He stated, "This composition of the lime or lemon acid, with wine and sugar, so administered, I esteem the most efficacious remedy for this disease, and greatly to exceed the simple lemon juice, or any other method in which it may be given." He added, "Upon repeated trials, I found that the virtues of lemon juice in this disease, exceeded those of green vegetables, and were much superior to that of wine by itself."[8]

As far as is known, Lind did not carry out any more controlled clinical trials at Hasler. Specifically, he never tested his rob. Moreover, he did not advocate forcefully for the provision of citrus juice or fresh fruits or vegetables to the Royal Navy. He stated, "The province has been mine to deliver precepts: the power is in others to execute." Modesty is not always a virtue. Despite Lind's brillant *Treatise*, the Admiralty took no action, and scurvy continued to plague the British navy for another forty years.

3

AN UNLIKELY HERO AND A PARTIAL VICTORY

Whenever you can, count.

—attributed to Francis Galton, 1924

CAPTAIN JAMES COOK: A POPULAR HERO

James Cook in many accounts gets credit for the conquest of scurvy, because, in stark contrast to Anson, he circumnavigated the globe without sustaining a single death from the disease.[1] Although Cook deserves credit for preserving the health of his men and improving the lives of British sailors, he set back efforts to prevent scurvy.[2]

Cook was born in 1728 to a farming family in Yorkshire, England. He felt the pull of the sea and apprenticed as a merchant seaman at the age of seventeen.[3] After rising through the ranks of the merchant marine, he joined the Royal Navy in 1755. He taught himself land surveying and astronomy and mapped the mouth of the Saint Lawrence River and the coast of Newfoundland. More than twenty years of experience as a ship's officer, expertise in navigation, and a desire to "go farther than any man has been before me" made him an excellent choice to lead a voyage of discovery.

On his first voyage around the world, Cook set out on August 25, 1768, from Plymouth on a refitted coal transport vessel renamed the *Endeavour*, carrying both sailors and scientists. The Admiralty and the Royal Society jointly sponsored the voyage, one of the first purely scientific naval expeditions in history.

Cook first sailed to Tahiti, where, because of cloud cover, he was unsuccessful in timing the transit of Venus across the sun (from which the distance of the earth from the sun could be calculated). He then turned south to search for the "Great Southern Continent," believed to exist to balance the land masses in the northern hemisphere. Pundits theorized that without a counterbalancing southern continent, the earth would wobble uncontrollably.

The *Endeavour* carried provisions for eighteen months at sea, including several putative antiscorbutics: malt (germinated and dried barley), sauerkraut, carrots, marmalade, mustard, saloup (a drink prepared from tree bark), portable soup, and Lind's rob of lemon and orange. Of these, only the sauerkraut had any value in preventing scurvy.

Of greater importance, the ship could land and reprovision frequently, as it was a time of relative peace. Cook paid attention to the health of his men and was assiduous in obtaining fresh fruits and greens whenever possible. The sailors were initially resistant to these unaccustomed foods until the officers made a show of eating them with gusto, and the sailors eventually came around. As a result, scurvy was never a major problem. After a voyage of two years, nine months, and seventeen days, Cook landed in England on June 12, 1771. Not one of his sailors had died of scurvy.

On his second voyage, Cook again set out to find the Southern Continent, this time with two ships, the *Resolution*, commanded by Cook, and the *Adventure*, captained by Tobias Furneaux. The ships departed Plymouth on July 13, 1772, and arrived at the Cape of Good Hope on October 30 with the crews in good health. They departed the cape on November 22, heading south and becoming the first European ships to cross the Antarctic Circle. Signs of scurvy appeared and were effectively treated, probably with sauerkraut. Once again, Lind's rob was tried and proved ineffective.

After 117 days at sea, the *Resolution* landed at New Zealand but had become separated from the *Adventure* in a dense fog. They rendezvoused on April 7 at Ships Cove. Like several prior expeditions, they inadvertently performed an experiment. The crew of the *Adventure* was suffering from scurvy, having not paid the same attention to diet as Cook's ship. The men of the *Adventure* rapidly recovered after eating fresh greens harvested in Ships Cove.

They repeated the experiment when the ships sailed from New Zealand to Tahiti. Three men on Cook's ship developed signs of scurvy and recovered with treatment. In contrast, thirty of Furneaux's men were ill with scurvy on arrival at Tahiti. Apparently, Furneaux was unable to learn from experience and provide fresh vegetables and fruits to his men.

The *Adventure* returned to England while Cook continued searching for the Southern Continent. He explored enough of the Southern Ocean to disprove its existence. Cook continued to land and reprovision with fresh greens and fruits whenever possible. He also served spruce tea to the crew at several stops. As a result, scurvy never was a serious problem on board the *Resolution*.

Cook landed in England on July 30, 1775. He received an award from the Royal Society for his observations concerning scurvy, and he wrote that malt wort (an infusion of malt prepared in boiling water, a byproduct of brewing beer) was "without doubt one of the best antiscorbutic sea medicines yet found out."[4] He did qualify his endorsement by saying that although malt wort would prevent scurvy, "I am not altogether of the opinion that it will cure it in an advanced state at sea." It is, in fact, worthless in treating scurvy. Cook may have been currying favor with the

Admiralty, which favored malt wort, in part because it was much cheaper than lemon juice.

Cook undertook a third voyage in 1776 to search for the Northwest Passage and was killed by indigenous Hawaiians on February 14, 1779. His crew returned to England on October 4, 1780, without discovering the Northwest Passage. The men had all remained in good health, there having been no signs of scurvy.

Cook proved that sailors could undertake long ocean voyages, even in inhospitable climates, without incurring the ravages of scurvy. However, he gleaned only minimal specific information about the disease. His shotgun approach to prevention and treatment—most importantly, frequently landing and restocking with fresh fruits and vegetables—although effective, gave no information about individual measures. He did try Lind's rob, and its failure was interpreted as a failure of citrus fruits in general. By the time of his death in 1779, Cook had achieved great credibility with the Admiralty, and his opinions influenced policy for more than a decade.

GILBERT BLANE: AN UNSUNG HERO

Although Captain Cook became famous, Gilbert Blane (also spelled Blayne) deserves most of the credit for conquering scurvy in the British Navy.[5] In the process, he founded the science of epidemiology.

Blane was born in 1749 to a prosperous merchant family in the southwest of Scotland. He went to the University of Edinburgh at age fourteen to study for the clergy, but switched to medicine, spending ten years at the university. He was elected president of the Students' Medical Society and garnered the good opinion of the faculty, both in the Faculty of Arts and the Medical School.

Armed with letters of recommendation from prominent Edinburgh scholars, he went to London and opened a medical practice.

He had a cold manner, later earning the nickname "Chilblaine." According to one description, "[he exhibited] a certain sanctified, devout, death-like expression of the countenance." Nevertheless, introduced into London society by a prominent physician, William Hunter, he established a successful medical practice.

One of his prominent patients was Sir George Rodney, admiral of the West Indian fleet, which had been racked with scurvy in previous action. In 1779, Rodney persuaded Blane to abandon his thriving practice and sail to the West Indies as his personal physician. Blane left no record of what led him to give up a comfortable life in London to go off to war.

He distinguished himself, both as a doctor and by helping on deck during battles. Rodney appointed him physician to the fleet. With the authority of this position, he required the ships' doctors and military hospitals in the region to submit monthly reports enumerating the illnesses and deaths among sailors of the fleet. These reports became the basis of his later writing and his efforts to affect naval policy.

Why did he believe that collecting such data would be useful? No one had ever done this before, and one can only imagine the grumblings of the doctors about unnecessary paperwork. Blane wrote that his primary motive was to distribute patients and resources efficiently among the hospitals in the region. He added: "These returns [reports] have served also . . . as a method of collecting a multitude of well-established facts, tending to ascertain the causes and course of disease."[6]

This insight is the foundation of epidemiology.

In 1781, between tours in the West Indies, Blane submitted a report, which he called a memorial, to the Admiralty. He noted that of twelve thousand men serving the fleet there had been sixteen hundred deaths, of which sixty had been due to enemy action and the remainder caused by infection and scurvy. He added:

Scurvy, one of the principal diseases with which seamen are afflicted, may be infallibly prevented, or cured, by vegetables and fruit, particularly oranges, lemons or limes. These might be supplied by employing one or more small vessels to collect them at different islands; policy, as well as humanity, concur in recommending it. Every fifty oranges or lemons might be considered as a hand to the fleet, inasmuch as the health, and perhaps the life, of a man would thereby be saved.[7]

In response, the Admiralty asked the Office for Sick and Hurt Seamen, the committee of physicians responsible for the health of the sailors, to render an opinion. Citing the experience of Captain Cook with Lind's rob, they advised that wort and sauerkraut were much more effective than fresh fruit and vegetables. The Admiralty accepted the advice of its panel of medical experts and ignored Blane. This is an early example of how one must be wary of advice given solely based on expert opinion.

However, Blane did not lose his interest in the health of sailors, nor did he surrender to the bureaucracy. Instead, he used the temporary truce to marshal his forces.

THE FIRST EPIDEMIOLOGIST

After service in the navy, Blane returned to London to resume his medical practice. He wrote an account of his experience in the West Indies, published in 1789 as *Observations on the Diseases of Seamen.*[8] It is an underrated work, possibly because it is one of the most boring books ever written. Nevertheless, it had a bigger impact on the history of scurvy and the history of medicine than Lind's *Treatise.*

Blane, like Lind, was a product of the Scottish Enlightenment, but, unlike Lind, he was not locked in to conventional theories. He thought that to make progress one had to "compare a great number of facts" and not rely on individual cases or preconceived hypotheses. He was an inveterate compiler of statistics. More than a third of his textbook consists of the tables of data he collected in the West Indies. The month-by-month and ship-by-ship health statistics, along with his analysis of these tables in turgid prose, make for tedious reading.

By studying patterns of disease incidence over time, including the ships and hospitals under his supervision, Blane made inferences about the causes of diseases and how to manage them effectively. These analyses of the incidence of disease across time and space form the basis of the science of epidemiology, and Blane was the first practitioner of that science.

He was convinced that Lind had proven that lemons, limes, and oranges were the most effective antiscorbutics available. To bolster that belief, he pointed to two ships that came into port after a battle and were able to purchase limes. The men on these ships recovered from scurvy, while the disease continued to increase on the other ships of the fleet, which did not have the benefit of fresh limes.

Blane drew other conclusions with practical consequences. First, he dismissed the importance of bad air—miasma—in causing scurvy: "It was remarked, that the men recovered faster on board than on shore; and it would appear that land air, merely as such, has no share in the cure of the scurvy, and that the benefit arises from the concomitant diet, cleanliness, and recreation."[9] He also noted that Lind's rob was ineffective. He blamed the heating of the juice for abolishing its antiscorbutic properties, although he did not say how he reached that conclusion.

Blane made several observations concerning the containment of contagious diseases on board ship. He went beyond Lind by using his observations to affect policy. Since hospitals at that time were incubators of infections, he kept sailors with scurvy on board ship for their treatment.

> It appears that only four men died of this disease [scurvy] in the whole fleet in the month of June, though there were so many ill of it; whereas it appears by the book of hospitals, that scorbutic men die there in much greater proportion, and chiefly in consequence of other diseases, particularly the flux [dysentery], which they catch by infection, or bring on by intemperance. It is farther in favour of this scheme, that great numbers of those sent on shore are lost by desertion. It is also a great saving to the Government, the expense not being a fourth part of what it would cost in a hospital.[10]

Given the state of medical care in the eighteenth century, the prevention of disease, not its treatment, was Blane's main concern.

> The prevention of diseases is as much deserving our attention as their cure, for the art of physic is at best fallible, and sickness, under the best medical management, is productive of great inconvenience, and is attended with more or less mortality. The means of prevention are also more within our power than those of cure, for it is more in human art to remove contagion, to alter a man's food and what air he is to breathe, than it is to produce any given change in the internal operations of the body. What we know concerning prevention is also more certain and satisfactory, in as much as it is easier to investigate the external causes that affect health than to develop the secret springs of the animal œconomy.[11]

His words still resonate in the twenty-first century.

BLANE'S VICTORY

Despite being rebuffed by the Admiralty Board in 1781 when he submitted his memorial, Blane won in the end. In 1795, he was appointed commissioner to the Board of the Sick and Wounded Sailors (formerly the Office for Sick and Hurt Seamen). He gained this appointment despite being, by all accounts, personally unlikeable. Moreover, he did not mince his words. In *Observations on the Diseases of Seamen*, he laid the blame for sailors' illnesses at the feet of the officers and the Admiralty.

> It was a saying of some of the ancients, that acute diseases were sent from heaven, whereas chronic diseases were of man's own creation. But I shall endeavor in the course of this work to evince, that, with regard to seamen at least, acute diseases are as much artificial as any others, being the offspring of mismanagement and neglect, with this difference, that they are imputable not so much to the misconduct of the sufferers themselves, as of those under whose protection they are placed.[12]

With his new authority, he persuaded the Lords of the Admiralty to provide, but not mandate, a dose of three-quarters of an ounce of lemon juice daily to each man. This was no longer Lind's ineffective rob, preserved by heating, but either fresh juice or juice preserved in alcohol. Although ships' captains did not all adopt the practice, there was a steep drop in the number of cases of scurvy in the British Navy.

This was the first example of the use of epidemiological data to guide government policy. It was a milestone in the history of medicine. Blane demonstrated that merely tabulating and analyzing when and where diseases occur can lead to better understanding of their cause and prevention, even when the disease is not contagious. The idea now seems obvious, but it had never been carried out on the scale and with the diligence of Gilbert Blane.

Most sources date the birth of the science of epidemiology to 1854, when the London physician John Snow analyzed a cholera epidemic.[13] Snow tracked all the cases in London and pinpointed the source of the epidemic to the Broad Street pump, a source of drinking water for a London neighborhood. This was before microorganisms had been discovered as the cause of infections, but Snow persuaded the authorities to remove the pump handle. Either as a result or coincidentally, the epidemic subsided, and the Broad Street pump became famous in the annals of medicine. Snow and Blane both deserve admiration, but Blane paved the way.

Thanks to Blane, after the death of more than a million sailors over three centuries, scurvy ceased to be the primary cause of death in the British Navy. In contrast to Lind, Blane analyzed his observations dispassionately, without trying to fit them into the theories he had been taught as a medical student. Although it took almost fifteen years, he used his insights to persuade the naval bureaucracy to improve the lives of its sailors.

Blane had a remarkable ability to ignore theory and false claims and see the problem of scurvy clearly. What gave him this objectivity? There are few clues. In an unfinished portrait, he looks like an accountant, and he had an accountant's love of numbers. He pored over the monthly reports from the fleet while he served the navy. His book spends many pages presenting the numbers and almost none discussing theories of disease.

It is equally remarkable that he was able to mold naval policy. According to one obituary, "The station attained may fairly be attributed to his talents and industry than to possession of external graces and artificial attractions."[14] He was persistent. He did not give up when snubbed by the Admiralty in 1781, but he remained interested in naval medicine and kept pressure on the bureaucrats. His prose was stilted, as was typical of his day, but clear and direct, and he knew more about his subject than anyone of his time. His victory was a triumph of careful science.

However, his success was not complete. Because the provision of antiscorbutics was not universal, scurvy did not disappear entirely from the Royal Navy. That required drastic action by the sailors themselves to give the final push to the Lords of the Admiralty.

THE ROYAL NAVY IN THE GEORGIAN ERA

Near-continuous warfare marked the reign of King George III, which extended from 1760 to 1820. Great Britain had been at war on multiple fronts: with its rebellious American colonies on one side and with European powers, primarily Napoleonic France, on the other. Additionally, Britain still had its overseas empire to defend. The navy was crucial to conflicts over much of the globe. Despite their importance to the country, British seamen not only shared the dangers and privations of all mariners during the Age of Sail, but they endured indignities unique to the Royal Navy.

As many as half of the sailors on some ships were prisoners in all but name, having been impressed into service. Impressment was the practice by which all able-bodied English mariners—merchant seamen and fishermen as well as former navy men—were subject to being drafted into the navy. Impressment was unique to Britain. Queen Anne instituted the practice in 1711 and subsequent monarchs continued it until the Napoleonic Wars ended in 1814.

Press gangs roamed the waterfront and received a bounty for each man they provided. The press gangs were authorized by law to take only experienced seamen, but in practice they took whomever they could find by any means available. They induced some men to serve voluntarily but took many by force, frequently while drunk. Many of the impressed men were old, infirm, and chronically malnourished. Others had just stepped ashore from their previous long voyages, still suffering the effects of months of hard labor and poor nutrition. When the need was desperate, men were taken directly from hospital wards.

Courts of law provided another source of recruits. Magistrates could sentence criminals to serve their time in the navy rather than jail. These criminals were not all illiterate thieves and muggers; some were white-collar criminals or men with unpaid debts, some were men with education, and some with experience running businesses. They formed a core group of potential leaders in the forecastle, where the ordinary seamen lived.

Common sailors still were not allowed to leave the ship when in port because of the risk of desertion. On board, they had few rights. Some of the officers were brutal. They resorted to flogging and keelhauling to maintain discipline. Captains had the authority to impose the death penalty, and sailors could be shot or thrown overboard for minor offenses.

Nevertheless, regulations required that the sailors at least have adequate food. The Victualling Board oversaw the system. Victualling officials, stationed in the ports, purchased supplies and distributed them to the ships' bursars. The Victualling Board gave exquisitely detailed specifications for the weekly ration:

> One pound avoirdupois clean, sweet, sound, well-bolted with a horse cloth, well-baked and well-conditioned biscuit; one gallon, wine measure, of beer; two pounds avoirdupois of beef killed and made up with salt in England, of a well-fed ox, for Sundays, Mondays, Tuesdays and Thursdays, or instead of beef, for two of these days one pound avoirdupois of bacon, or salted English pork, or a well-fed hog, and a pint of pease (Winchester measure) therewith; and for Wednesdays, Fridays and Saturdays, every man, besides the aforesaid allowance of bread and beer, to have by the day the eighth part of a full-size North Sea cod of 24 inches long, or the sixth part of a haberdine 22 inches long, or a quarter part of the same if but 16 inches long; or a pound avoirdupois of a well-savoured Poor

John [dried fish cake], together with two ounces of butter, and four ounces of Suffolk cheese, or two-thirds of that weight of Cheshire.[15]

Although deficient in vitamins, this diet would have provided sufficient caloric intake to support hard physical labor and included a necessary mixture of carbohydrates, protein, and lipids.

However, in Georgian England, every level of the provisioning process was corrupt. By the time the provisions reached the sailors, they were reduced in both quantity and quality. Kickbacks, adulteration, short weights, and theft of supplies were rife. The exact constitution of the actual diet is unknown but was certainly less than prescribed by the Victualling Board.

Ships were usually provisioned for six months, and the food spoiled over time. The biscuits grew moldy and worm-infested, the beer spoiled, and the cheese and butter became rancid. Since ordinary seamen could not go ashore, they had no opportunity to supplement this diet.

The British military strategy depended on the naval blockade of enemy ports and constant patrolling of the Channel and North Sea. Ships were frequently at sea for months at a time, some only a few miles off the coast of Britain. Although it was possible to re-provision ships periodically using tenders, as suggested by Gilbert Blane, this was not done. The authorities saw this as unnecessary, as they already had supplied the ships with ample provisions and were unwilling to incur the added expense of providing fresh food.

Pay was the sailors' number-one grievance. The continuous warfare had bankrupted the Crown, and lack of money was at the root of many of the navy's problems. Although the army had received a recent pay raise, sailors had not received an increase in more than thirty years despite a rise in the cost of living. Some crews had not been paid for more than two years. Their only hope of reward was a share of the booty from captured enemy vessels.

Faced with these conditions, the sailors rebelled. The men of the Channel Fleet took the opportunity in spring 1797 when they anchored at England's main naval base at Portsmouth.

THE SPITHEAD MUTINY[16]

At the end of March 1797 and after a month at sea, the sixteen large warships and more than twenty smaller craft of the Channel Fleet sailed into the Solent, the narrow channel between Portsmouth and the Isle of Wight on the south coast of England. They came to refit and reprovision. The country was on edge, expecting an armada of French and Dutch ships to attack any day. The navy, and especially the Channel Fleet, had to be at peak readiness to defend the islands.

They moored at Spithead, an anchorage two to three miles offshore from Portsmouth. In the distance the sailors could see the town bustling with wartime activity. The Solent was crammed with ships. In addition to the Channel Fleet, dozens of vessels filled the docks and shipyards of the huge naval base next to the town.

As Easter was only two weeks away, most officers took the opportunity to go ashore and enjoy town life or visit their families, leaving the sailors on board the closely packed ships with little supervision. Anchored in calm waters, they had a respite from their labors and could talk among themselves out of earshot of the few officers still on board. The spirit of the French Revolution was in the air, and this was an opportunity for the men to press their many grievances. If they stuck together, they had leverage, as the nation was depending on them to fend off the expected invasion. To coordinate their plans, the leaders rowed back and forth between ships, usually under the cover of darkness. They agreed to take decisive action.

On Easter Sunday the order came to sail. The men refused. In a coordinated action, they banished the officers to shore, established self-government on the ships, and refused to put to sea until the Admiralty addressed their demands. They presented their petition to the Admiralty in the form of letters signed by representatives of each ship. After asking for a raise in pay, the sailors listed their demands for better conditions on board:

> We, your petitioners, beg that your Lordships will take into consideration the grievance of which we complain, and now lay before you.
>
> First, That our provisions be raised to the weight of sixteen ounces to the pound, and of a better quality; and that our measures may be the same as those used in the commercial code of this country.
>
> Secondly, That your petitioners request your Honours will be pleased to observe, there should be no flour served while we are in any port whatever, under the command of the British flag; and also, that there might be granted a sufficient quantity of vegetables of such kind as may be most plentiful in the ports to which we go; which we grievously complain and lay under want of.
>
> Thirdly, That your lordships will be pleased seriously to look into the state of the sick on board His Majesty's ships, and that they may be better attended to, and that they may have the use of such necessaries as are allowed for them in time of sickness; and that these necessaries be not on any account embezzled.[17]

Other demands included shore leave while in port, continuation of pay for men injured in the course of duty, and removal of brutal officers.

The naval hierarchy first ignored these demands and then obstinately refused to address them, threatening to charge the perpetrators with the capital offense of mutiny. The sailors remained steadfast. After a month, with an invasion expected any day and Parliament growing restless, the Admiralty had no choice but to address the grievances and King George to grant the mutineers a pardon.

The mutiny spread unevenly to other ships, but these did not achieve the success of the Spithead fleet. Many of those participants were executed or exiled to Australia. Because of the prevalence of scurvy on the ships ferrying prisoners to New South Wales, exile to Australia was a death sentence for many. The Admiralty did not expend resources providing fresh vegetables or lemon juice to prisoners.

The Spithead Mutiny had far-reaching consequences for the Royal Navy, even though its success was limited to the Channel Fleet. The pay for common seamen was raised, and they were allowed shore leave while in port. Officers with a record of brutality were removed, and the officer corps became more professional. The Admiralty began to pay attention to living conditions on board their ships. More attention was given to keeping the ships clean, and the navy provided the men dry clothing and bath soap. Fresh vegetables were added to the sailors' diet, and the authorities enforced the resolution of 1795 to provide citrus juice to all the ships, a policy that to that point had been applied inconsistently.

Gilbert Blane and the ordinary seamen had won. Scurvy virtually disappeared from the Royal Navy for more than a half century. But, remarkably, not permanently.

4

STEPS FORWARD
AND BACK

*[I]t is no easy matter to root out old prejudices, or
to overturn opinions which have acquired an establishment by time, custom, and great authority.*

—James Lind, preface to
Treatise on the Scurvy, 1753

SCURVY AT SEA

After Gilbert Blane's reforms of 1795, admissions of sailors with scurvy to the Haslar Naval Hospital plummeted.[1] Between 1806 and 1810, there were only two cases.[2] This drop occurred for two reasons. First, improvements in diet greatly reduced the incidence of scurvy in the navy. Second, Blane instituted the practice of treating ill sailors on board their ships, rather than in hospitals, unless they required major surgery. Since germs had not yet been discovered, Blane did not understand how infections spread, but his numbers told him that they killed men in hospitals.

Despite the decline of scurvy in the navy, merchant seamen suffered from the disease for another half century.[3] In 1854, Parliament made a half-hearted attempt to address the problem by mandating that merchant seamen be given lime juice but without specifying how much. The law was not enforced, and men continued to develop the disease. In response to pressure from the Seamen's Hospital Society and articles in the *Times* of London, in 1867 Parliament passed the Merchant Shipping Amendment Act, which specified both the quantity and quality of lime juice to be provided to each mariner. Moreover, the act strengthened enforcement of the law and stiffened penalties for breaking it.

The 1867 law produced a decline, but not the complete disappearance, of scurvy cases on merchant vessels. The seamen of the London docks were served by a hospital ship until 1870, when the hospital was moved to dry land. Prior to 1867, there was an average of about ninety cases of scurvy per year, compared to about thirty per year subsequently.[4] Only after 1885, by which time the steamship had taken over the merchant service, did the disease became rare.

The quality of the lime juice was another issue. At the end of the eighteenth century, the navy stopped using Lind's rob and began using the juice of lemons or limes either mixed with rum or briefly brought to a boil and bottled under a layer of olive oil. Omitting the prolonged heating of the juice better preserved its vitamin C and contributed to the decline in scurvy in the navy after 1790.

In 1860 the navy made another change. It had previously obtained lemons from Malta, Spain, and Italy. When warfare disrupted the supply lines in the Mediterranean, the navy switched to limes from the Caribbean. To the navy, it was all lime juice. It made no distinction between lemons and limes, nor did it consider the origin of the fruit.

It also began using a new preparation of lime juice. In 1867, Lachlan Rose patented a method for preserving lime juice in a concentrated sugar solution rather than alcohol.[5] Rose's Lime Juice later became popular as an ingredient in cocktails, but his first major customer was the navy. He opened a factory in Leith, Scotland, near the wharves and became a major supplier. He expanded his business to London in 1875. Rose also obtained his limes from the West Indies. Those limes contained much less vitamin C than Mediterranean lemons, leading, as we shall see, to more scurvy and more confusion.

SCURVY ON LAND

Although scurvy grew increasingly rare among sailors during the nineteenth century, it continued to occur on land. Scurvy was common during the California gold rush of 1849 and 1850 and the Crimean War of 1854 to 1856, among the among military prisoners in the American Civil War of 1861 to 1865, and among French settlers of the Saint Lawrence region of Canada.[6] As late as World War I, British soldiers in the Middle East suffered an outbreak of scurvy.[7]

However, in normal times, Great Britain and northern Europe did not suffer from scurvy, even in the winter when most people went several months without eating fresh vegetables. The potato, introduced to Europe from the New World in the seventeenth century, played a key role. Although the potato is not as rich in vitamin C as citrus fruit or leafy green vegetables, a single potato per day provides enough vitamin C to prevent scurvy.

Potatoes required less land than grain crops and could be stored to eat in the winter. By the nineteenth century, they had become a staple, especially in Ireland and Scotland, where the typical winter diet of peasant farmers consisted of potatoes, milk, and bread with butter or molasses. Irish tenant farmers raised other crops, but they were cash crops for the export market, sold to pay the rent on the land. Potatoes fed the farmers and their families. Men doing hard labor ate on average more than nine pounds of potatoes per day.[8]

Prisons brought the importance of the potato to light. The Milbank Penitentiary in London, completed in 1816, was organized as a model of humane rehabilitation of criminals.[9] Milbank provided hygienic conditions and a good diet so that prisoners, although denied their freedom, would not be deprived of their health. Rehabilitation occurred through isolation, hard labor, and religious instruction.

In July 1822, the prison directors decided that they were coddling the prisoners by giving them a better diet than they would enjoy in their poverty-stricken lives on the outside. To redress this excess of humanitarianism, they removed potatoes from the diet. Within three months, scurvy appeared. By March 1823, half of the prisoners—but none of the officers, civilian staff, or kitchen workers—were suffering from the disease.

The physicians charged with investigating the epidemic persuaded the authorities to give the prisoners oranges, and the scurvy outbreak rapidly subsided. However, they did not give oranges to military prisoners in the same prison. They remained on a diet lacking potatoes and fresh vegetables and continued to suffer from scurvy.

Twenty years later, in 1843, William Baly, physician to the Milbank Penitentiary, published a paper analyzing the epidemiology of scurvy in British prisons and advocated potatoes as a preventative.[10] He noted that Milbank's military prisoners had suffered from scurvy even after cooked vegetables—likely boiled until they were mush—were added to their weekly ration of soup. After potatoes were added in January 1842, no more cases occurred.

Baly described other British prisons in which scurvy disappeared when potatoes were included in the diet. He pointed out that the antiscorbutic properties of raw potatoes had been long known and that light cooking does not destroy their benefit.

Administrators approved because potatoes were cheaper than oranges or lemons.

The Potato Famine of 1845 to 1848 reinforced Baly's conclusions. In the summer and fall of 1845, a fungus infection rapidly spread across Europe and destroyed the potato crop.[11] The next three winters, famine devastated Ireland. Waves of typhus, relapsing fever, cholera, and dysentery killed those whose malnourished bodies grew too weak to resist infections. Those who escaped fatal infections died of multiple organ failure or from exposure, forced to sleep in ditches after being evicted from their houses when they could not pay the rent.

In Ireland, out of a population of about eight million people, probably a million died.[12] Another million emigrated to escape the famine, resulting in a 25 percent loss of the population.

Some impoverished families were able to obtain enough calories to stave off starvation. Relief efforts by the British government and by private agencies tried to lessen the devastation by importing corn meal from America and operating soup kitchens to feed the destitute. These efforts saved some from starvation, but cases of scurvy began to appear, since corn meal and weak soup, frequently boiled for hours, provided calories but no vitamin C.

Most physicians of the time had never seen a case of scurvy and were confused by the disease. Initially, some physicians misdiagnosed the condition as purpura hemorrhagica because the dark spots on the skin due to scurvy resemble those seen in bleeding disorders. As the disease became common throughout the British Isles, physicians recognized it as scurvy.[13]

The experience in prisons and during the Potato Famine underlines two points. First, importing the potato from the New World two centuries earlier had spared northern Europe from yearly winter epidemics of scurvy when potatoes provided the only source of vitamin C for millions of the poor. When the potato crop failed, there was no staple food to take its place, and scurvy appeared.

Second, starvation kills before a person can develop scurvy. Only those who could obtain enough calories from corn meal, bread, or grains to stave off gross malnutrition developed scurvy. It takes at least two months—but typically longer—before the body's stores of vitamin C are used up. Starvation kills more rapidly. The failure to develop scurvy by people totally deprived of food was one factor that prevented physicians from recognizing that scurvy is a nutritional deficiency disease.

SCURVY IN INFANTS

The story of scurvy in infants revolves around milk and begins with the Siege of Paris.[14] During the Franco-Prussian War, Prussian troops rapidly advanced across France and laid siege to Paris from mid-August 1870 through the following January. With provisions to the city cut off for four-and-a-half months, milk and fresh vegetables became unobtainable. Scurvy, often combined with other nutritional deficiencies, became common by the end of the siege, especially among prisoners, who were the first to suffer the food shortage.

Without milk, young children died of malnutrition. Physicians tried to save the children by feeding them other sources of protein, but the effort failed. The experience spurred research into the nutritional properties of milk and attempts to develop a substitute. These efforts paved the way to an eventual understanding of vitamin deficiencies.

Human breast milk provides sufficient vitamin C for an infant, but breast feeding became unpopular during the nineteenth century.

Wealthy women hired wet nurses. Those less well-off turned to milk substitutes and fed their infants commercial powders dissolved in water, condensed milk, or diluted cow's milk. The powders provided protein but lacked vitamins. Cow's milk, even fresh and undiluted, is a poor source of vitamin C.

Hospitals began to see children with a new disease marked by severe pains in the legs. The doctors did not recognize this as scurvy. Confusing the issue, the infants frequently also had rickets, a disease caused by vitamin D deficiency that produces bony deformities. It was common among children who ate a limited diet and had little exposure to sunlight in the urban streets darkened by air pollution. Since both rickets and the new disease affected the long bones of the legs, physicians called the condition "acute rickets."

Thomas Barlow, a pediatric physician, saw his first case in 1875, when he was a registrar (resident physician) at the Hospital for Sick Children, Great Ormond Street, London.[15] He assisted a surgeon, Thomas Smith, in the clinical description and postmortem examination of a twenty-three-month-old girl with episodes of swollen, painful legs.

The child screamed when her legs were touched or even when a nurse approached her bed. Because the girl did not have swollen, bleeding gums, Barlow dismissed the diagnosis of scurvy. She was "rickety." However, since rickets is not painful, Barlow knew that rickets could not explain the child's severe leg pain.

The girl died suddenly while still hospitalized. The remarkable postmortem findings included hemorrhages around the bones and a peculiar separation of the growth areas at the ends of the long bones from the shaft. These findings were later to become crucial clues to recognizing scurvy in laboratory animals.

Barlow studied more of these cases and came to recognize the disease as infantile scurvy. In 1883 he published a paper entitled "Cases Described as 'Acute Rickets' Which Are Probably a Combination of Scurvy and Rickets, the Scurvy Being an Essential, and the Rickets a Variable, Element."[16] The paper is a model of clinical and pathological description and analysis. He summarized thirty-one cases, of which eleven had been under his personal care. He describes a typical case, that of a fifteen-month-old boy:

> During the first six weeks he was said to be a vigorous child. For that period, he had his mother's milk, then it entirely failed, and from that time until when I saw him he had been quite deprived of fresh food. At first his diet consisted of Robinson's grits and Swiss milk, then of baked flour, then of Nestle's food, then of Robb's biscuits, then of Liebig's extracts, and finally of Swiss milk and saccharated lime water. . . .
>
> The child had been able to sit up well and stand with assistance at thirteen months old. Five weeks ago he ceased to do either, and then it was noticed that the left leg was swollen especially about the ankle. At this time also he became very peevish and would shriek if he were touched, and often even if he were approached. . . .
>
> The child has an excessively pale, sallow complexion. . . . He has cut his two lower incisors; the gums are natural with the exception of a minute erosion in the upper gum opposite the cutting edge of one of the lower teeth. The boy is continually moaning and when approached he screams and still more when he is touched. . . . Both the left thigh and the leg are slightly swollen so that the contour of the limb is different from natural, assuming in the thigh a rather cylindrical shape.

Barlow performed autopsies on three cases and concluded that the pain and swelling of the legs were caused by the hemorrhage surrounding the bones. Lind had reported similar hemorrhages in adult sailors, and Barlow made the connection. He also recognized that scurvy did not cause swollen, bleeding gums in infants prior to teething. He therefore diagnosed infantile scurvy.

Barlow recognized that scurvy was a nutritional disease. He described his treatment of a typical case. Besides applying warm compresses to the legs to relieve the pain, two teaspoons of orange juice daily were added to the diet.

> In three days, there was a notable change in the child. . . . The left lower limb was less tense and less tender, and the right leg was better. . . . After this the improvement was progressive. . . . Within eight weeks from the date when first seen, the boy, whenever allowed, would get upon his knees and could stand with a little support; he was of a ruddy colour, and his skin and muscles had become quite firm.

The rapid improvement with a change in diet was further evidence that the disease was scurvy. Infantile scurvy became known as Barlow's disease.[17]

Early in the twentieth century, Alfred Hess and Mildred Fish, two investigators from the New York City Department of Health, observed a small outbreak of infantile scurvy and studied what foods best returned the children to health.[18] Before refrigeration, urban institutions could not conveniently provide fresh, whole milk, as it spoiled during transport and storage. In 1912, the Hebrew Infant Asylum in New York City began to feed cows' milk pasteurized by heating to 145 degrees Fahrenheit to infants. The administrators of the institution had been told that heating the milk did not affect its properties. Since they were now giving the infants a new wholesome source of nutrition, they felt justified in removing orange juice from the children's diet. Several children developed scurvy as a result.

Hess and Fish investigated the outbreak on behalf of the Department of Health and observed firsthand the children's recovery when given either orange juice or raw milk. Those given orange juice recovered rapidly. With raw milk, recovery was slow and inconsistent. They also tried carrots, which were ineffective, and mashed potatoes, which were effective but not as dramatically so as orange juice. Hess and Fish concluded:

> Raw cow's milk must not . . . be considered as having potent antiscorbutic properties. Its effect cannot be compared to the miraculous change which is brought about by giving orange juice. This is especially striking when we take into consideration the small amount of orange juice necessary to bring about a cure and compare it with the large amount of raw milk which is given. Raw milk, however, contains sufficient of the essential substances to prevent the development of scurvy.

Hinting at events to come, they went on to say, "In this connection we must mention the very interesting and suggestive studies of Funk. This author has coined the word 'vitamines' for substances which are essential to the health and life of the body."

Caimer Funk played a pivotal role in the evolution of this story. Moreover, the work of Hess and Fish motivated a biochemist in Pittsburgh, Charles Glen King, to become interested in nutrition and to become one of the discoverers of vitamin C.

SCURVY AT THE ENDS OF THE EARTH

During the late nineteenth and early twentieth centuries, countries raced to be first to the poles. Equivalent of the Cold War space race, the contest captured the public's attention much as space exploration has captured ours. When polar explorers died, the public response was like the reaction to the explosion of the space shuttle *Challenger*. Scurvy among polar explorers brought the disease to the attention of the public as well as more confusion.

In the early 1600s, the Dutch sent expeditions into the far north, not to reach the pole but to find a northern route to the East Indies. These attempts initially met with failure when explorers died during the long winters. However, some expeditions survived for up to two years, subsisting almost exclusively on fresh meat, mainly eaten raw. Fresh meat is a poor source of vitamin C, but as the only component of their diet, it contains barely enough to ward off scurvy.

One tale of survival in the Arctic dates from 1630, when eight men from an English whaling expedition were stranded on the east coast of Greenland during the winter.[19] They survived by killing game, including polar bears. They soon developed scurvy. When they ate the polar bear liver, their skin peeled due vitamin A toxicity, but their scurvy improved. In the spring, all were rescued and recovered their health. They had eaten just the right amount of polar bear liver. It is a good source of vitamin C, but a better source of vitamin A. Vitamin A toxicity can cause the skin to slough off entirely and be fatal.

The eighteenth-century Dutch physician Friedrich Bachstrom told another remarkable story of survival, perhaps apocryphal.[20] Bachstrom, ahead of his time, was convinced that scurvy was a dietary deficiency disease and that the only certain cure was fresh vegetables. To support his contention, he told of a sailor marooned on Greenland. The man was so debilitated by scurvy that he could not stand. All he could do was crawl on the ground and graze on the grass like a wild animal. Eating grass soon restored him to health and he survived.

Arctic exploration began generating confusion about scurvy in 1875, when Captain George Nares led a widely publicized British expedition to the Arctic.[21] He left England with two ships, the *Alert*

and the *Discovery*, and 121 crewmen. The *Alert* sailed up the west coast of Greenland, where it spent the winter trapped in the ice well north of the arctic circle.

In April 1876, sixteen men under Commander Albert Markham set out from the ship on sledges to push as far north as possible. They had taken their daily doses of lime juice on board ship. It was standard naval issue, prepared from limes from Monserrat in the West Indies. The juice may have been pumped through copper pipes during bottling. Copper is a potent catalyst for the oxidation of ascorbic acid, destroying its antiscorbutic activity.

Because of the difficulty of thawing liquids in temperatures well below freezing and the belief that the men had left in a well-nourished state, they did not take lime juice on the sledges. Within two weeks, the first signs of scurvy appeared. The men pushed north for another week before turning back. By the time a search party from the ship met them in early June, scurvy had disabled all but four men, and one had died.

On returning to the *Alert*, they found that many other crew members were also suffering from scurvy and that the *Discovery* was experiencing a similar occurrence. The expedition made it back to England, but four men died and sixty more developed scurvy.

Parliament thought these events sufficiently important to establish a commission of inquiry. The commission compared the Nares expedition to one in the 1850s, which experienced no scurvy after twenty-seven months in the arctic and lost only three of sixty-five men after being trapped in the ice for three winters. They took what the navy called lime juice but was in fact prepared from Mediterranean lemons, not West Indian limes. That was the crucial difference.

The commission, not understanding that difference, tried to decipher why the Nares expedition fared so much worse than the earlier one, evaluating various factors including diet. The failure of the lime juice confused the committee. They did not consider the origin of the fruit and concluded that the cause of the outbreak of scurvy was the leaders' failure to provide lime juice to the sledging parties. They glossed over the occurrence of scurvy among the men who remained on the ship, downing their daily doses.

Controversy ensued. Noted scientists of the Royal Geographical Society refuted the report. They questioned the effectiveness of lime juice and cited other instances in which it had failed to prevent scurvy. One academic physician, C. R. Markham, the cousin of Commander Albert Markham, the leader of the sledging expedition, termed the doctrine of dietary antiscorbutics such as lime juice "the last remnant of an obsolete physiology."[22]

The reaction to the committee's report showed that not everyone believed that scurvy was caused by faulty nutrition. Captain Robert Falcon Scott, the most famous British polar explorer, was a prominent dissenter. From 1901 to 1904, he led an expedition to Antarctica aboard the ship *Discovery*. Scurvy was a recurrent problem while the ship remained trapped in ice during two winters.

A physician member of his crew, Edward A. Wilson, published an account of the journey in 1905 in the *British Medical Journal*.[23] He held to the ptomaine theory, a theory that arose after the discovery of disease-causing microorganisms in the late nineteenth century, which were invoked to explain all diseases. Hence the idea arose that bacteria contaminating tinned food produced a toxin, ptomaine, which caused scurvy. It surprised Wilson when scurvy struck the men even though "every tin of meat [was] examined by sight and smell."

Wilson changed the diet to minimize the use of canned food. Six days a week the explorers ate fresh seal meat briefly fried in butter. On Thursdays they ate the tinned meats; the men called these "scurvy days."[24] "On Sundays we had seal's liver for breakfast, the most favorite dish of all."[25]

Although they did not realize it, seal liver is rich in vitamin C, and scurvy rapidly disappeared. Wilson continued to believe that the scurvy had been caused by the tinned meats and cured by eating fresh meat, not understanding the importance of the seal liver. Based on his observations, he claimed that "fresh meat will rapidly cure scurvy without either lime juice or fresh vegetables." This is true only if the fresh meat is raw or barely cooked liver. Wilson's paper illustrates a scientific maxim: correlation does not prove causation.

The story of Scott's final and ill-fated *Terra Nova* Expedition of 1910 to 1913 has been told many times.[26] He and Wilson continued to dismiss the importance of diet in causing scurvy. Lime juice was available in the base camp, but Scott did not require his men to take it. The five men in the final sledging party to the south pole went without fresh foods for more than four months. They ate biscuit, bacon, chocolate, and pemmican (a mix of dried meat and fat) and took no vegetables or lime juice.

On their way back from the pole, all five perished. The following spring, a search party found three of the bodies, including Scott and Wilson, in their tent. They did not mention any signs of scurvy, nor did Scott in his diary. It would have been remarkable if Scott's party did not suffer from the disease after a prolonged period on a diet lacking vitamin C. Some have suggested that the truth was covered up to avoid casting aspersions on the conduct of the expedition. The search party buried the bodies in their sleeping bags where they found them. We probably will never know the truth.

SCURVY IN COMMITTEE

At the end of the nineteenth century, a disease misnamed "ship beriberi" attracted the attention of Norwegian naval authorities, who also favored the ptomaine theory.[27] Beriberi is a disease that affects the peripheral nervous system, and it is now known to be caused by a lack of thiamin, a B vitamin. It had become endemic in South Asia during the nineteenth century. Sailors involved in trade between Europe and Asia developed a disease called ship beriberi, which was thought to be a form of that Asian disease. However, unlike Asian beriberi, ship beriberi did not cause degeneration of the nerves, and some sufferers had gum and skin changes typical of scurvy. It was likely a disease of multiple vitamin deficiencies, though primarily scurvy.

At the end of the nineteenth century, the Norwegian navy changed ships' rations and supplied its ships with canned food, both meats and vegetables. Canned meat could be preserved for long voyages and was more palatable than salted meat. When Norwegian sailors developed ship beriberi, the Norwegian naval authorities charged a committee with investigating the outbreak.

In 1902, the committee issued its report, giving more support to the ptomaine theory. With no bacteriological evidence to support their conclusion, they blamed the disease on contaminated tins of food. They were partially right in blaming the canned food. It had been sterilized by heating, and the committee had no way of understanding the effect of heat in promoting the oxidation of ascorbic acid in the vegetables. Hence, lacking a better explanation, they understandably invoked bacterial contamination.

Four hundred years after the crew of Vasco da Gama had discovered the ability of oranges to cure scurvy, experts remained confused about its cause. Miasma had been dismissed as a cause, but three hypotheses continued to vie for acceptance. First, the belief that scurvy was an infection resonated with the new germ theory of disease, and this belief was a favorite of Russian physicians. Second, the ptomaine theory held that scurvy was an indirect effect of germs caused by a toxin in food tainted by bacterial contamination. And finally, many authorities, especially British naval physicians, believed that scurvy was a dietary deficiency. The nature of the deficiency remained a mystery.

Although Blane had gotten it right more than one hundred years earlier, many medical authorities were questioning his wisdom at the end of the nineteenth century. The failure of lime juice to protect polar explorers and the ptomaine theory had muddied the waters. But with hard work and some remarkable luck, the waters would soon clear.

THE CHEMISTS
TAKE OVER:
THE DISCOVERY
OF VITAMINS

A DIFFERENT KIND
OF NUTRIENT

Scientific knowledge advances haltingly and is
stimulated by contention and doubt.

—Claude Lévi-Strauss,
The Raw and the Cooked 1969

THE BIRTH OF A SCIENCE OF NUTRITION

There was no eureka moment when a single brilliant researcher discovered vitamins. Nobody deserves the sole credit. Rather, incremental advances punctuated distinct lines of research, and strokes of good luck alternated with bewildering wrong turns. Some investigators wore blinders formed by outmoded and inappropriate disease models. However, persistence and luck led to the eventual recognition of what we now take for granted: nutrients in foods present in only trace amounts are essential for health.

The systematic investigation of human nutrition began in 1815 when the French Academy formed the Gelatin Commission to find ways to feed the growing numbers of urban poor.[1] The Gelatin Commission started with the belief that there are only two basic nutritional requirements: enough total calories to provide energy and adequate protein to repair tissues. If true, the poor could subsist on bread to provide the calories and any source of protein. The commission hoped that gelatin could provide that protein cheaply.

Gelatin is the product of boiling animal parts that are composed mainly of connective tissue—skin, tendons, hooves, and bones—in water or a weak acid. The boiling extracts the collagen protein from the tissues and makes it digestible. The Gelatin Commission hoped that after slaughterhouses stripped carcasses of meat to feed the well-off, gelatin prepared from the offal could feed the poor.

The chair of the Gelatin Commission, the noted physiologist Francois Magendie, performed a series of groundbreaking experiments between 1816 and 1841. He fed dogs restricted diets, including sugar alone, bread alone, gelatin alone, and bread and gelatin. Despite having as much food as they wanted, all the dogs fed these restricted diets lost weight and died. This surprised and disappointed the commission, as it meant that its supposition was wrong: bread and gelatin by themselves could not sustain life.

Even though he did not discover a cheap way to feed the poor, Magendie introduced two innovations that paved the way to the discovery of vitamins. First, he used animals as models of human physiology. He assumed that processes essential to life, such as nutrition, are similar in all mammals. Previously, animals had been used in nutrition research only in animal husbandry investigations. For example, investigators compared the amount of milk produced by corn-fed dairy cows to those fed barley. The Gelatin Commission wanted to understand the nutrition of people, not farm animals.

The second innovation was the use of simplified diets. Rather than feeding complete natural foods, such as grains or meat, Magendie fed the dogs refined sugar and gelatin. The animal experiments proceeded in parallel with advances in chemistry, mainly in Germany, that produced a more detailed understanding of the composition of foods. Researchers could exploit this knowledge to test chemically defined ingredients. The Gelatin Commission failed to feed the poor, but it began a line of investigation to define minimum nutritional requirements.

The experiments were simple in principle: feed the purified food components to groups of animals, weigh the animals periodically, observe their general health, and see how long they live. In practice, many details required attention. Did the animals in fact eat the food, or was it so unappetizing that they starved themselves? Did they eat enough at first but then tire of eating the same food day after day? Even if they ate the food, was it absorbed from the intestine or merely excreted in feces? How was the food prepared? Was it fresh or had it been stored for weeks? Had it been heated to prevent bacterial growth? Rigorous experiments demanded labor and resources. Researchers paid variable attention to the crucial details.

A MEDICAL STUDENT MAKES A BIG DISCOVERY

From the time of the Gelatin Commission until the turn of the twentieth century, the infant science of nutrition grew into a major scientific effort and attracted public attention. Continuing the line of research initiated by Magendie, investigators found that animals required a mixture of protein, fat, carbohydrates, inorganic salts, and water to survive. Chemists measured the energy content of foods, expressed in calories. Physiologists exploited advances in chemistry to study diets of purified nutrients to define minimum nutritional requirements.

Nikolai Lunin, a Russian medical student working at the University of Basel in Switzerland, made a major advance.[2] He knew that mice could live indefinitely fed only whole cows' milk. Lunin's innovation was to separate the milk into its major known components: casein (the main milk protein), milk fat, lactose (the sugar in milk), and inorganic salts. He then mixed them back together and fed the mixture to mice. The mice all died in a few weeks. In the process of purifying the major components of milk, Lunin had lost something that was present in only small amounts but that his mice required to live. In his 1881 paper, he concluded, "a natural food such as milk must therefore contain besides those known principal ingredients small quantities of unknown substances essential to life."

This was a revolutionary insight. Lunin was the first to recognize the existence of essential nutrients present in natural foods but not present in their purified major ingredients. If anyone deserves sole credit for the discovery of vitamins, it is Lunin. His was the first mention in the medical literature that a disease could be caused by the lack of a nutrient other than carbohydrates, protein, and fats. We now term this class of substance *micronutrients* and Lunin deserves credit for their discovery.

The work of this Russian medical student attracted little attention, even though he worked in the laboratory of the noted physiologist Gustav von Bunge, who referred to the work in his widely read textbook. For almost three decades, no one searched for what Lunin had lost when he split the milk into separate components.

SCIENCE IN THE JUNGLE: BERIBERI

The next breakthrough came not from university laboratories in Europe but from a military outpost in Indonesia. During the nineteenth century, tropical beriberi had become common in regions where white rice was the staple of the diet: Southeast Asia, India, Japan, and the Philippines.[3] The disease, now known to result from a deficiency of the B vitamin thiamin, produced degeneration of the peripheral nerves, beginning in the legs and causing weakness, muscle wasting, and numbness. As the disease progressed, the legs and abdomen swelled, and eventually the victim died of heart failure. By the late nineteenth century, beriberi had become a cause of death equal to infectious diseases in South Asia.

A kernel of rice has three layers: an outer, indigestible husk, several layers of cells that form a thin silverskin, or bran (the pericarp), and the white core (the endosperm). The germ, the actual seed, is at the base of the core. Grinding off the husk, leaving the core and a layer of silverskin produces brown rice. Polished white rice requires grinding off the silverskin, an outer layer of cells of the core and the germ. The silverskin and the germ contain thiamin. The white core is almost entirely starch.

Most people preferred white to brown rice but polishing rice by hand with a mortar and pestle was laborious. Consequently, white rice was a luxury through the eighteenth century. With the invention of steam-powered milling machines in the early nineteenth century, white rice became cheap enough to replace brown rice as the main component of the diet in most Asian countries. As the Medical Research Council of Britain stated, beriberi became common in "rice-eating districts of the East when they had been invaded by milling machinery from the West."[4]

Yet no one understood this at the time. After Louis Pasteur and Robert Koch discovered disease-causing bacteria in the 1870s, the germ theory dominated medical thinking, and investigators began looking for the germ that caused every mysterious disease. The prevailing wisdom was that beriberi occurred when people in some tropical countries became infected with a bacterium or parasite that produced a neurotoxin.

The first evidence that this was wrong came from a Japanese naval physician, Kanehiro Takaki.[5] On returning to Japan in 1880 from medical studies in London, he noted that beriberi afflicted only common seamen, whose diet was mainly white rice supplemented with a little fish. Ships' officers, who ate a more varied diet, never developed the disease.

Takaki understood that this was not the pattern of an infectious disease, which would have spread among all the men on a ship regardless of rank. He surmised that beriberi was the result of an inadequate diet, although he thought the deficiency was of protein.

Despite being wrong about what was lacking in the diet, he convinced the Imperial Japanese Navy to supplement the diet of common sailors with barley. This simple change in diet eliminated beriberi from the Japanese navy. As an example of the refusal of human beings to alter thinking in the face of evidence, most Japanese physicians continued to believe that beriberi was caused by an infection. The Japanese army refused to add barley to the diet, and the disease continued to kill its soldiers by the thousands.[6] However, Takaki won naval honors and the nickname "Barley Baron."

Doctors in Europe had not heard of Takaki. In 1886, Holland, following the conventional wisdom, dispatched Christiaan Eijkman to the island of Java to find the microbe that was causing the outbreak of beriberi in its Southeast Asian colonies.[7] Eijkman was perfect for the job. He was a physician and physiologist who had worked with Robert Koch, the founder of microbiology.

In Java, Eijkman toiled in a bare-bones laboratory attached to a military hospital on the outskirts of Djakarta, then called Batavia. He conducted his initial experiments with rabbits, but he soon switched to chickens, probably because they were cheaper. He injected groups of chickens with the blood of beriberi patients, expecting it to transfer the causative microbe. He left control groups uninjected for comparison.

He made no progress, until suddenly in 1889, all his chickens, whether injected with patients' blood or not, developed leg weakness and died. He performed postmortem examinations and found that the chickens suffered the same nerve degeneration as humans with beriberi. Being conservative, Eijkman was unwilling to assume that his chickens had developed beriberi, so he named the avian disease *polyneuritis gallinarum*, meaning inflammation of the nerves of chickens. Nonetheless, he had in fact produced an experimental model of tropical beriberi.

Sticking to his belief that beriberi was an infectious disease, he started over with new chickens and investigated the possibility that an airborne organism had killed the first group. After making no progress for a few months, he got another surprise. The disease mysteriously disappeared as suddenly as it had appeared. His chickens stopped developing leg weakness and remained healthy.

On further investigation, he discovered that the original practice in the laboratory had been to feed the chickens uncooked wholegrain rice, the cheapest form of rice. The chickens had their own milling mechanisms in their gizzards to remove the husks. However, the assistant responsible for the care of the chickens wanted to save even more money and arranged for the hospital cook to donate the polished white rice left over from feeding the patients. Shortly after this change in diet, the chickens developed nerve degeneration. A few months later, a new cook arrived and "refused to allow military rice to be taken for civilian chickens."[8] Returning to a diet of whole-grain rice, the chickens remained healthy.

Unwilling to abandon the germ theory, Eijkman tried to fit the data to the theory. He proposed that a toxin-producing microbe infected the rice kernel and that a substance in the bran was an antitoxin. He still thought that diseases had to be caused by some active agent—a microbe or a toxin. A negative factor, such as a nutritional deficiency, was alien to his way of thinking.

Despite his incorrect assumptions, he performed the correct experiments. He verified that a diet of only white rice led to the peripheral nerve degeneration and that the condition could be reversed by adding back the silverskin. He also showed that the material in the silverskin that protected the chickens was soluble in water.

Eijkman continued to work in Java until 1896, when he became ill with malaria and returned to Europe to assume a prestigious professorship at the University of Utrecht, the preeminent medical school of that time. He left Java still clinging to the germ theory and believing that beriberi was caused by a microbe.

Gerrit Grijns, a physiologist who had never worked directly with Eijkman, took over the work in Java and approached the problem with an open mind. He repeated Eijkman's experiments with the same results. He also showed that carbohydrate diets other than white rice, including milk sugar and potato flour, led to nerve degeneration. These results made the infected rice theory unlikely.

Grijns recognized that beriberi was a nutritional disease and concluded that there was some essential, water-soluble nutrient in the bran that was missing in the milled grains. He finally convinced Eijkman of this obvious conclusion, which Grijns published in 1901.[9]

Working in an isolated hospital and using the most basic tools, these two researchers verified Lunin's conclusion: even when provided adequate calories and basic nutrients, a deficiency of some unknown substance could lead to a fatal disease. Thus, they verified

Lunin's novel disease model and established another milestone in the history of nutrition research.

Eijkman won the Nobel Prize in Physiology and Medicine in 1929. He was too ill to travel to Stockholm for the award ceremony and sent a lengthy letter of acceptance as a substitute for a Nobel Prize address delivered in person.[10] He did not mention his colleague Grijns, without whose insight Eijkman may have continued looking for a germ and never won any prizes.

PARADIGM SHIFT: A DIFFERENT KIND OF DISEASE

Thomas Kuhn in his influential essay *The Structure of Scientific Revolutions*, published in 1962, described the uneven nature of scientific progress.[11] He used the word *paradigm* to refer to an accepted model of how nature functions. A paradigm includes both the theories and the experimental methods that are applied to expand and test the model. Isaac Newton's laws of motion are an example of such a paradigm, and for more than two hundred years it guided astronomers in their observations and explanations of the movement of the planets and other heavenly bodies.

Kuhn pointed out that science does not progress smoothly. Rather, periods of "normal science," guided by a reigning paradigm, are punctuated by sudden jumps, which he called paradigm shifts, revolutions in scientific thought that result in a new model and fresh ways of viewing a problem.* Albert Einstein's theory of relativity was a paradigm shift in understanding the laws of motion. It led astronomers to analyze the movement of heavenly bodies in new ways, to ask new questions, and to design novel ways of observing the skies.

---

*Although Kuhn's essay was a major contribution to the history of science, it did a disservice to the language. Now almost every scientific publication, no matter how trivial, promises a paradigm shift.

Between revolutions, "normal science" proceeds as investigators apply the reigning paradigm and perform experiments that fill the gaps of knowledge about old problems and explore new questions. Normal science proceeds until observations are made that the old paradigm cannot explain, requiring a new way of thinking, a new paradigm.

The work of Eijkman and Grijns is a prime example of a paradigm shift. Eijkman started out doing normal science in Java. He applied the reigning paradigm, the germ theory, to design and interpret experiments with his chickens. He did this until things happened to his chickens that the germ theory could not explain. He was eventually forced, by his observations and by the arguments of Grijns, to accept a new disease model, that of a nutritional explanation for beriberi. And this new model permitted breakthroughs in understanding other diseases. The first was scurvy.

SCURVY FINALLY EXPLAINED

In 1907, Norwegian naval authorities commissioned two researchers, Axel Holst and Theodor Frølich, to investigate the cause of ship beriberi, which was afflicting Norwegian sailors returning from the East Indies. Although the disease was not the same as tropical beriberi, Holst and Frølich set out with the assumption that it was. They first verified the findings of Eijkman and Grijns.[12] They fed milled grain to pigeons and produced the same degeneration of peripheral nerves as Eijkman had produced in chickens.

They then tried to create a model of tropical beriberi in mammals by feeding guinea pigs a diet of milled grains similar to that which had produced nerve degeneration in pigeons. They never explained why they chose guinea pigs, but it was one of the strokes of luck that punctuated vitamin research. Guinea pigs, like humans, cannot synthesize vitamin C.

The cereal diet produced a fatal disease in the guinea pigs, but it was not beriberi. There was no degeneration of the peripheral nerves. Moreover, unlike avian beriberi, it made no difference if the guinea pigs ate white or brown rice. As long they ate only cereal grains, whether it was rice, barley, or oats, they developed the disease.

Holst and Frølich were not the first to produce this disease in guinea pigs. Theobald Smith, a physician and bacteriologist working for the U.S. Department of Agriculture, had done so in 1895 by feeding the animals a diet of oats and bran.[13] The guinea pigs died with bleeding into deep tissues. If he supplemented their diet with grass, clover, or cabbage, the guinea pigs remained healthy.

Smith did not know what was killing his guinea pigs. His main interest was microbes that infect pigs, so he never followed up his observations. He went on to a distinguished career in microbiology. By discovering that insects could transmit infectious diseases, he became one of the first American medical scientists to achieve international recognition. But he missed his chance to discover the cause of scurvy.

Holst and Frølich did not waste their opportunity. They made the astute observation that their guinea pigs had abnormalities of the bones identical to those described in 1883 by Thomas Barlow in fatal cases of infantile scurvy. Furthermore, when the guinea pigs' diet of grains was supplemented with lemon juice, apples, potatoes, or cabbage, the guinea pigs either did not develop the bone disease or showed only slight abnormalities.

Holst and Frølich correctly concluded that they had produced scurvy in the guinea pigs and that scurvy was a nutritional deficiency disease caused by a "one-sided diet" lacking in some essential substance or substances. They also found that cabbage heated in an autoclave for a half hour in pressurized steam at 120 degrees Celsius was less effective in protecting the animals than if only heated in unpressurized boiling water. Hence, they concluded that strong heating could inactivate the antiscorbutic factor, an observation that explained why citrus juice preserved by prolonged boiling and the canned foods fed to the Norwegian sailors were ineffective in preventing scurvy.

With their 1907 publication, Holst and Frølich at long last showed that scurvy was a disease caused by a restricted diet. The paradigm shift of Eijkman and Grijns bore fruit in understanding another disease. With the work of Holst and Frølich, suddenly much of the confusion concerning scurvy was swept away.

However, when a paradigm shift occurs, investigators who have invested their careers in the previous way of thinking may resist challenges to that model. Charles Darwin's theory of evolution met strong resistance from many naturalists, and some molecular biologists resisted the concept that a protein, termed a prion, could cause infections even though it did not carry genetic information. Similarly, some researchers continued to question the conclusion of Holst and Frølich and continued looking for other explanations.[14]

Rats, which can synthesize vitamin C from glucose, do not develop scurvy if given the same diet that produces the disease in guinea pigs. This observation contributed to the confusion about the cause of scurvy. Some investigators refused to believe that there were fundamental differences in basic nutritional requirements among mammals, and they continued to seek other explanations. Meanwhile, more flexible minds explored the new paradigm.

VITAMINES AND VITAMINS

In 1912, an English scientist, Frederick Gowland Hopkins, published a paper that essentially repeated Lunin's work. But unlike Lunin, Hopkins garnered attention from his fellow scientists.[15]

Hopkins was a member of the British academic establishment. He was on the faculty of Cambridge University, having been educated at the University of London and Guys Hospital. Although the term "biochemist" did not exist at the time, he was in fact one of the first. He was interested in how cells generate their energy. This led him to study the effects of diet on the growth of young rats.

Possibly following up on the inability to find a milk substitute to feed infants during the Siege of Paris, Hopkins studied young, growing rats. Hopkins found that young rats fed a basic, chemically defined diet similar to that used by Lunin remained healthy for only two weeks, at which point they began to lose weight and died by about four weeks. When he supplemented the diet with a small amount of cows' milk, less than half a teaspoon per day (less than 4 percent of the animals' food), the young animals grew normally and survived to adulthood.

Like Lunin, Hopkins concluded that there was a substance or substances in milk essential for normal growth and survival of young animals, but it was required in only small amounts. In his 1912 paper, he termed these "accessory factors in normal dietaries," which he later modified to "accessory food factors."

The terminology became a matter of contention. Hopkins had a rival, Casimer Funk, a Polish-born chemist.[16] Whereas Hopkins was a member of the Oxbridge elite, Funk was an outsider. As a Jew, he had to leave Poland to pursue his university education in Switzerland. Subsequently, whether as a cause or a result of his feelings of alienation, he continued to move around.

Funk became interested in nutrition while studying in Berlin with Emil Fischer, a pioneer of organic chemistry. There, Funk investigated the nutritional properties of proteins, a line of research that later led others to identify essential amino acids, those that could not be synthesized by mammals and are required in the diet.

In 1911, after moving to the Lister Institute in London, Funk took up beriberi research. Following on the work of Eijkman and

Grijns, he set out to isolate the antiberiberi substance. Although he was unsuccessful in these efforts, he wrote a review article published in 1912, the same year as Hopkins's landmark paper. In that paper, he proposed his deficiency theory of disease.[17] He reasoned that diseases such as pellagra and rickets, beriberi and scurvy, were caused by a lack of a substance present in only small amounts in certain foods. In his review, Funk coined the term *vitamine* to refer to these substances. (He likely pronounced it *veetameen*.)

Funk's coinage was a combination of *vita*, Latin for life, and *amine*, meaning a nitrogen-containing compound. He later termed the antiberiberi factor *vitamine B*. Funk thought these substances were all nitrogen containing—that is, amines—hence the final *e*. He did not have strong evidence for that assertion, so his term *vitamine* was controversial among nutrition researchers. It quickly caught on with the public and continued in common use until 1920, when through the diplomatic suggestion of another nutrition researcher, Jack Cecil Drummond, it was shortened to *vitamin*.[18]

F. Gowland Hopkins continued to take issue with this term, with or without a final *e*, and championed his terminology, "accessory food factors." It is easy to understand why "vitamin" won out. The argument over terminology was a surrogate for claiming credit for the discovery of vitamins, which neither of them deserved. If anyone merited that distinction, it was Lunin, based on his work decades earlier. Nonetheless, a Russian medical student did not command the same respect as a Cambridge professor.

THE DEMANDS OF WAR

The outbreak of World War I gave nutrition research increased urgency. Beriberi devastated both British and Turkish troops during the disastrous Gallipoli campaign. British troops in India and the Middle East developed scurvy and got no help from the same preparation of lime juice that had failed the polar explorers. The military needed foods that were easy to transport, could tolerate both hot and cold climates, and could nourish the Allied soldiers fighting around the world.

On the British home front, men left their jobs to join the military, leaving women to step into leadership positions in science.[19] Harriette Chick was one of those women.[20] She was born in 1875 into an upper-middle-class London family with a legacy of independent women who had built the family's lace business. Her father was a conservative Protestant fundamentalist, but he enrolled Harriette in a high school that was ahead of its time in teaching mathematics and science to girls. Harriette was a star student and, along with four of her sisters, was among only the second generation of women to attend a British university.

She studied botany at University College London and, after her undergraduate degree, did microbiology research. In 1904 she earned a doctor of science degree from University of London for work on algae in polluted waters. Subsequently, over the vehement opposition of two members of the all-male faculty, she won a fellowship to work at the Lister Institute for Preventive Medicine in London, the major scientific research institution in Great Britain prior to World War I.

She first worked on disinfectants and related protein chemistry, continuing her interest in microbiology. With the onset of war, almost the entire male staff at the institute left to join the army. To contribute to the war effort, Chick dropped her own research and moved to producing antisera to diagnose and treat battlefield infections. The director of the Lister Institute, Charles Martin, had gone to the Greek island of Lemnos to serve as pathologist in a military hospital. When he saw troops returning from Gallipoli suffering from beriberi, he asked Harriette Chick to hand over the routine antisera work to others and take charge of nutrition research at the institute. Casimer Funk had moved to the United States, opening the door for Harriette Chick to take over.

This was a turning point in Harriette Chick's life. She took to the new assignment with gusto and became one of the most important nutrition researchers of the twentieth century. With colleagues Margaret Hume and Ruth K. Skelton, Chick carried out a series of meticulous feeding experiments to quantify the ability of various foods to prevent beriberi and scurvy.[21] Unlike other investigators, these women cared for the animals themselves, feeding them by hand to ensure that they ate exactly what the experimental protocol prescribed.

They made several crucial observations. First, that lime juice prepared from limes from the West Indies, even when fresh, had only one-quarter the antiscorbutic potency of lemon juice prepared from Mediterranean lemons. They tested samples of the navy's preserved lime juice and found that it lacked any antiscorbutic activity. They showed that the antiscorbutic value of other foods also decreased over time with storage and verified that it was destroyed by strong heating. They also quantified the antiscorbutic activity of a variety of fresh fruits and vegetables.

Alice Henderson, also at the Lister Institute, complemented these laboratory findings with a historical study of the use of antiscorbutics in the Royal Navy.[22] Her papers, published in 1919 after the war had ended, documented that scurvy had all but disappeared among British sailors when the navy provided them juice of Maltese lemons. However, in the 1860s, when the military switched to juice prepared from West Indian limes, scurvy reappeared among polar explorers and soldiers in the Middle East. The findings of Chick and Hume showed why that switch resulted in scurvy persisting more than four hundred years after the crew of Vasco da Gama recognized the curative effects of fresh fruit.

Most of the women who pursued science during World War I married and gave up their careers when the war ended. During that era, married women were virtually excluded from professional positions in Britain. Harriette Chick remained single and continued her work at the Lister Institute, going on to a distinguished career in nutrition research.

Her most important work concerned rickets, the disease that caused weak and deformed bones in children and is now known to result from a deficiency of vitamin D. She, along with colleagues from the Lister Institute, traveled to Vienna in 1919 and 1920 to study malnutrition in that war-ravaged city. Working in a children's hospital, she showed that giving children cod liver oil or exposing them to sunlight or to an ultraviolet lamp prevented rickets.[23] She continued to do nutrition research until her retirement in 1945. After that, she remained on the board of the Lister Institute and active in the Nutrition Society, to which she gave a talk two weeks before her hundredth birthday.

Chick's discoveries concerning rickets were on par with the work of Eijkman in their impact on public health. Eijkman received the Nobel Prize, whereas Harriette Chick received little recognition beyond her professional circle. A pediatrician would most likely give a blank stare if asked about Harriette Chick.

HOLDOUTS

Some researchers could not accept the paradigm shift brought on by the nutritional deficiency hypothesis. One holdout was the prominent American physiologist, Elmer V. McCollum. He worked at the University of Wisconsin and later assumed the prestigious chair of the newly formed Department of Chemical Hygiene at Johns Hopkins University.

McCollum reported his discovery of the fat-soluble factor A required for the growth of young rats in a 1913 paper.[24] Turning to scurvy research, he and his coworker W. Pitz produced the disease in guinea pigs but not in rats, and they did not believe that animal species would differ markedly in their nutritional requirements. Inexplicably, they also failed to prevent scurvy in guinea pigs fed a diet of oats and milk, a regimen that had been successful for Holst and Frølich.[25]

They found that their scorbutic guinea pigs had impacted fecal material in their colons and concluded in a 1918 publication that a toxin present in the impacted feces caused scurvy, concluding that the benefit of citrus juices resulted from the laxative properties of citric acid. However, when McCollum learned of the meticulous work of Chick and Hume, he realized his error and recanted.

THE VITAMIN ALPHABET

In 1913, the year after the publications of Hopkins and Funk, E. V. McCollum and Marguerite Davis at the University of Wisconsin published a paper showing that the substance or substances in milk that supported the growth of young rats in Hopkins's experiments was present in the butterfat and was lipid soluble.[26] The essential material was also found in egg yolk. Since it was lipid soluble, it was distinct from the water-soluble antiberiberi substance.

Hence, there was a fat-soluble substance in butterfat and egg yolk needed to support the growth of young animals, and there was a water-soluble substance found in the bran of grains (and also found in yeast) needed to prevent beriberi. These were initially termed *fat-soluble A* and *water-soluble B*, respectively, and later became *vitamin A* and *vitamin B*. Neither was the antiscorbutic substance.[27]

In 1919, Jack Cecil Drummond made the definitive pronouncement on the cause of scurvy.[28] He and others had found that rats can grow on a diet of purified protein, butterfat, salts, and carbohydrates supplemented with yeast extract. Drummond showed that they grow slightly better if also supplemented with orange juice. Although the relation of these findings to scurvy is unclear, it convinced the remaining doubters that there was a "water-soluble C," the antiscorbutic factor. Finally, after more than four hundred years of uncertainty, scurvy became universally accepted as a nutritional deficiency disease. And water-soluble C became vitamin C.

But these factors were more complicated than initially thought. The lipid-soluble material turned out to be a mixture of two substances. One retained the name vitamin A and the other became vitamin D, since C was already taken. And the material in bran and yeast was a mixture of several substances required for health. They all kept the name vitamin B but were given numbers in addition. The antiberiberi substance kept its first place in line as vitamin B1. The other B vitamins were associated with other diseases, mainly skin diseases and anemia, and were numbered two through twelve. A few dropped by the wayside, hence there are a total of eight B vitamins. Subsequently two more fat-soluble vitamins were discovered and termed vitamin E and vitamin K.

Vitamin C kept its place in the alphabet.

NOMENCLATURE AND NOBEL PRIZES

Through the first three decades of the twentieth century, the new science of nutrition was a major focus of medical research. In the *Journal of Biological Chemistry* and the *Biochemical Journal*, the primary journals of biochemistry, a sizable portion of articles during those years dealt with nutrition. It was then, as it is now, of great interest to the public. During World War I, it had taken on importance for the defense of nations. Hence, understanding vitamins offered researchers academic prestige and public recognition. It engaged the egos of many investigators and did not always bring out their best.

In 1906, the Cambridge professor F. Gowland Hopkins gave a speech to the Royal Society of Chemistry, mainly dealing with mundane administrative matters. Toward the end of the speech, he switched topics and predicted that rickets and scurvy were diseases caused by a lack of dietary "minimal qualitative factors."[29] This speech turned out to be important in the subsequent battles over credit for the discovery of vitamins. It is unclear at what point Hopkins turned to vitamin research in his own laboratory. He stated that his 1912 paper reported results "I obtained as far back as 1906–1907." Funk took issue with this claim and said that prior to a conversation he had with Hopkins during the 1910 Christmas break, Hopkins had performed no experiments with milk. Funk implied that he had given Hopkins the idea for the nutrition experiments.

In his 1912 paper, Hopkins renamed the unknown substances in milk essential to the growth of rats "accessory food factors." He admitted that he did not know the nature of these factors, but he offered three hypotheses. One was that they were substances that were not required for energy but were required for tissue repair. Second, they were substances required to synthesize other molecules, such as enzymes, that are essential for life. Third, and closest to the truth, they were substances that either served as catalysts or were required for the catalytic function of enzymes.

In the same year, Casimer Funk coined the term *vitamine*, later shortened to vitamin. Subsequently, he and Hopkins engaged in a long-running dispute. A Hungarian Jew outsider clashed against a bastion of the British academic establishment. The outsider won the battle over nomenclature, but he lost the war. The 1929 Nobel Prize in Physiology and Medicine went to Hopkins and Eijkman.

Eijkman was deserving, as he had explained the first disease, tropical beriberi, caused by vitamin deficiency and produced a paradigm shift. Hopkins's contributions were less original. He essentially repeated the work of Nikolai Lunin. Other investigators, notably McCollum and Davis and Osborne and Mendel in the United States, did similar work around the same time as Hopkins.

Casimer Funk was nominated for the Nobel Prize, but he did not share it. Although in his review paper he had enunciated the concept of nutritional deficiency disease, his original experimental work was of minor importance, and he never could purify the antiberiberi factor. Inexplicably, Holst and Frølich did not win the Nobel Prize. Hopkins at least acknowledged them, along with Lunin, in his Nobel Prize lecture. However, he devoted most of his speech to disputing the priority claims of Funk.

To give Hopkins his due, he was instrumental in founding biochemistry as an academic discipline. He was to make another important contribution to vitamin C research. He mentored another scientist with a big ego, Albert Szent-Gyorgyi.

6

THE VITAMIN HUNTERS

*Truth is a will o' the wisp that can only be caught
in the net of glory-scorning experiment.*

—Paul de Kruif, *The Microbe Hunters*, 1926

In science, being first counts for everything. Being second to make
a discovery carries little more prestige than copying a great work
of art. The early twentieth century was the first time that a scientist
could garner enough glory to make racing to be first worthwhile.
Vitamins, a hot topic to scientists and to the public, provided the
arena for well-publicized contests. Being the first to purify and
chemically characterize a vitamin would bring academic prestige
and public recognition. The effort consequently attracted investigators with big ambitions and egos to match.

THE TASK OF THE BIOCHEMIST

By the 1920s, the scientific community had accepted that scurvy
resulted from a dietary deficiency of a substance essential for life

but required in minuscule amounts. "Vitamin" had prevailed as the name for that kind of nutrient, but vitamins were defined only by their biological activity. There was a substance that prevented beriberi and another that prevented scurvy. Their chemical properties were understood only in broad strokes, and their molecular structures were unknown.

Vitamin C was the substance in citrus juice and many other fruits and vegetables that prevented scurvy. It was water soluble, acidic, and more stable in acidic than basic solutions. It lost potency—that is, its ability to prevent scurvy—with heating or when exposed to air. But exactly what it was—what atoms it contained, how those atoms were arranged into a molecule, how it could be commercially manufactured—remained a mystery.

One of the prominent investigators of the time, Sylvester Solomon Zilva (who understandably went by S. S. Zilva) said, "The ground was now prepared for the task of the biochemist."[1] For the biochemists to do their job properly, they needed pure vitamin C, meaning crystals of the compound. And getting those crystals proved to be a challenge. The general strategy was to start with plant material—lemon juice, for example—and try to progressively remove contaminating substances. At each step, the researchers tested what remained to verify that it still contained vitamin C. Although lemon juice was the richest known source of vitamin C, it had two disadvantages. First, it contains far greater quantities of contaminating sugars than of the vitamin. The chemical properties of the vitamin were so similar to the sugars that separating them proved difficult. Second, the vitamin was unstable in lemon juice and rapidly lost its potency when exposed to air.

In addition, the only method available for detecting vitamin C, termed a biological assay, was to test the ability of material to prevent or cure scurvy in animals, most conveniently, guinea pigs. The investigator placed groups of guinea pigs on a scurvy-inducing diet, usually a cereal grain known to contain the antiberiberi vitamin B supplemented with butterfat or milk fat to provide vitamins A and D. The test substance was added to that diet, and the investigator observed its ability to either prevent or cure scurvy.

For a quantitative assay, multiple groups of animals had to be used, each group receiving a different amount of the test substance. The assay was time consuming, as it took three to four weeks for guinea pigs to develop scurvy. In addition, it was imprecise and expensive, but the investigators had no alternative until a chemical assay could be developed, and a chemical assay required knowing what the substance was.

Two laboratories, one in Europe and one in the United States, attacked the problem head-on. S. S. Zilva started working on vitamin C in 1918 at the Lister Institute in London.[2] He worked for fifteen years and came close to purifying vitamin C from lemon juice, but he never quite reached his goal. He removed the solids and precipitated the citric acid from the juice, retaining the vitamin C activity in the remaining solution. Working at low temperatures and progressively precipitating various contaminants out of the solution, he obtained almost pure vitamin C. However, the antiscorbutic substance remained unstable, and he could never crystalize the pure compound.

Charles Glen King at the University of Pittsburgh also started with lemon juice. He achieved the goal but did not get the credit. That story requires a detour through Hungary.

A LUCKY MAN

The man who did receive the credit for purifying vitamin C was the biochemist Albert Szent-Gyorgyi.[3] He was born in Budapest to an upper-middle-class family with an aristocratic heritage—his full name was Albert Imre Szent-Gyorgyi von Nagyrapolt. Early in his career, he kept the aristocratic *von*, signing himself Albert von

Szent-Gyorgyi, but he soon gave it up. The pronunciation of his name is close to the English Saint Georgy, and he was content with being called that or merely "Saint George" by those intimidated by the Hungarian spelling.

His colorful and peripatetic life later included working as an undercover agent during World War II to help Hungary avoid the ravages of the Nazi occupation.

> On the one hand, my inner story is exceedingly simple, if not indeed dull: my life has been devoted to science and my only real ambition has been to contribute to it and live up to its standards. In complete contradiction to this, the external course has been rather bumpy.[4]

Despite the bumps in his personal life, his professional life was to benefit repeatedly from extraordinary good fortune.

He began his scientific career working in the anatomy laboratory of his uncle, Mihaly Lenhossek, the most prominent Hungarian medical researcher of the time. Lenhossek wanted his nephew to study diseases of the rectum. Albert was unenthusiastic about the rectum, but he was sufficiently interested in medicine to earn his medical degree from the University of Budapest. He graduated just as World War I broke out and was immediately was conscripted into military service and sent to the front lines. He apparently did not enjoy being a soldier, since he shot himself in the arm to escape the army. Following the war, when Czechoslovakia invaded Hungary in 1919, he packed up and, with wife and daughter in tow, moved from city to city across Europe, working in one temporary research position after another.

He eventually secured a position as a chemist in a laboratory at Groningen in the Netherlands, where he developed an interest in biological oxidation in both plants and animals. He started with the seemingly trivial question of why some fruits and vegetables turn brown when their cut flesh is exposed to air, whereas others retain their pale color. Examining this latter group, he found that they contained a reducing substance, something that protected their flesh from oxidation. He could not know it at the time, but the unknown antioxidant was vitamin C. Based on entirely erroneous speculation that this substance was a hormone, he also looked for the same activity in the adrenal gland, where he found it in large quantities. He later mused, "I must admit that most of the new observations I made were based on wrong theories." And, one might add, lucky guesses.

When the head of his laboratory in Groningen died, Szent-Gyorgyi once again had to move on. And once again he landed on his feet, securing a fellowship at Cambridge University with the help of the prominent professor of physiology, Frederick Gowland Hopkins, who became his mentor.

While in Cambridge, Szent-Gyorgyi succeeded in crystallizing a small amount of his reducing substance from oranges as well as from lemons and cabbages. The substance was a carbohydrate that had properties resembling a sugar. When he submitted his paper to the *Biochemical Journal*, he proposed the name "Ignose," since he was ignorant of what it was and *-ose* is the ending for the chemical names of sugars. The editor of the journal did not share Szent-Gyorgyi's sense of humor; he also rejected "Godnose." Szent-Gyorgyi and the editor finally agreed on "hexuronic acid," as it contained six carbon atoms and was acidic.[5]

Hopkins sent some of the material to S. S. Zilva at the Lister Institute. Inexplicably, Zilva reported that the material was not vitamin C. Zilva never explained the basis for this conclusion, but coming from a highly regarded expert, his pronouncement carried a great deal of weight.

Szent-Gyorgyi could purify hexuronic acid in large quantities only from adrenal glands, which were not readily available in Cambridge. So in 1929 he once again pulled up stakes and went to the Mayo Clinic in Minnesota, where he had access to the slaughterhouses in St. Paul. He crystalized twenty-five grams of the material and sent half to Walter N. Haworth, a carbohydrate chemist in

Birmingham, England, to determine its chemical structure. Haworth could not complete the structural chemistry with the quantity of compound provided. Possibly deterred by Zilva's assertion that hexuronic acid was not vitamin C, Szent-Gyorgyi put the project on the back burner.

In 1931, Szent-Gyorgyi accepted the chair of medicinal chemistry at the University of Szeged in his hometown in Hungary, taking with him his remaining supply, about one gram, of hexuronic acid. Soon after he arrived, Joseph Svirbely, an American of Hungarian heritage, walked unannounced into the laboratory in Szeged and offered his services to determine if hexuronic acid was vitamin C. Svirbely had just earned his PhD in the laboratory of Charles Glen King, the nutrition researcher at the University of Pittsburgh and Szent-Gyorgyi's rival. Svirbely came to Hungary on an Institute of International Education fellowship.

One of Szent-Gyorgyi's strokes of luck was that this young man appeared out of the blue. Svirbely had gained experience in the biological assay for vitamin C in King's laboratory. Like S. S. Zilva in London, King was trying to purify vitamin C from lemon juice. He and Svirbely had published a paper reporting their early results using the assay.[6] Svirbely, exploiting his experience with the assay, used Szent-Gyorgyi's last gram of hexuronic acid to test whether hexuronic acid was the antiscorbutic vitamin.

He divided guinea pigs into four groups. One group, the "positive controls," received a scurvy-inducing diet of rolled oats, bran, butterfat, salts, and a small amount of milk powder. They all died of scurvy after an average of twenty-six days. Another group received the same basal, scurvy-inducing diet as the positive controls but were supplemented with one milliliter per day of lemon juice, an amount known to prevent scurvy. They remained healthy. A third group received a standard laboratory diet, and they also remained healthy. The crucial "test animals" received the basal, scurvy-inducing diet plus one milligram of hexuronic acid per day.

The question was whether pure hexuronic acid would prevent scurvy just as well as lemon juice. The answer was that it did. Hexuronic acid was vitamin C.

Svirbely and Szent-Gyorgyi laid out their evidence.[7]

- A tiny amount of pure hexuronic acid, roughly twice the amount in a milliliter of lemon juice, was as good as the lemon juice itself or a standard laboratory diet in preventing scurvy.
- Hexuronic acid was contained in a variety of foods in proportion to their content of vitamin C.
- The chemical properties of hexuronic acid coincided with the known properties of vitamin C.
- The chemist Norman Haworth attested to the purity of Szent-Gyorgyi's hexuronic acid.

S. S. Zilva, disappointed to have been beaten to the finish line, continued to publicly demur, but the conclusion was inescapable. Szent-Gyorgyi's hexuronic acid was vitamin C.[8]

Szent-Gyorgyi later said that he had long suspected that hexuronic acid was vitamin C but had never tried to test it. He enjoyed the chemistry laboratory but avoided animal research. Also, he found vitamins "theoretically uninteresting." He wrote dismissively, "What one has to eat is the first concern of the chef, not the scientist."[9] But it was vitamin C that propelled him to scientific prominence.

A CONTESTED RACE

While Szent-Gyorgyi and Svirbely were working in Hungary, C. Glen King in Pittsburgh was pursuing the same goal: to show that hexuronic acid was vitamin C. The race ended in a dead heat.

King was a biochemist who became interested in vitamins after reading the work of Hess and Fish on infantile scurvy in New York

City. He was a careful and well-trained researcher and an expert in the bioassay of vitamin C. He purified hexuronic acid from lemon juice, but he also may have gotten some that was prepared from animal adrenal glands by Szent-Gyorgyi's former colleague Edward C. Kendall at the Mayo Clinic. King found that hexuronic acid was vitamin C at about the same time as Szent-Gyorgyi.

The virtual tie created a public fight over who was first. American supporters of King went up against European colleagues of Szent-Gyorgyi. Pittsburgh was the Silicon Valley of its time, creating new industries that were disrupting old ways of doing business. The upstarts in the blue-collar city in the hills of Pennsylvania felt that they were being disrespected in favor of a Hungarian aristocrat backed by Oxbridge insiders.

The fight came down to two versions of the detailed chronology of events. King had written to his trainee Svirbely in mid-March saying that he had tentatively concluded that hexuronic acid was vitamin C but was holding off on publication until he was sure of the purity of his hexuronic acid preparation. He also may have wanted to check out the claim of another investigator that vitamin C was a derivative of narcotine, a compound found in poppies and chemically unrelated to hexuronic acid. In early March, shortly before King mailed his letter, Svirbely had written from Hungary to his former mentor telling him that he and Szent-Gyorgyi had shown unequivocally that hexuronic acid was vitamin C. The two letters may have crossed in the transatlantic mail.

Svirbely's letter apparently prodded King into submitting a report to the American journal *Science* to establish priority. His paper, published in the issue dated April 1, 1932, announced that hexuronic acid and vitamin C were identical. Thus, King in fact published first. However, his publication in *Science* provided no experimental details.[10] He submitted a complete manuscript with details of his experiments to the *Journal of Biological Chemistry* a month later, on May 9, 1932.[11]

Szent-Gyorgyi, upon seeing the letter from King, dashed off a report to *Nature*, the European equivalent of *Science*.[12] The paper appeared in the April 16, 1932, issue and stated the same conclusion as King—that hexuronic acid is vitamin C—but with the important difference that it included Svirbely's guinea pig bioassay results to back up the claim. Thus Szent-Gyorgyi and Svirbely were the first, by less than a month, to publish unequivocal evidence that hexuronic acid was the antiscorbutic factor. The scientific community, at least in Europe, assumed that King had not wrapped up all his loose ends by the time he sent the letter to *Science* and credited Szent-Gyorgyi with being first.

King disputed this version of the chronology. He said that he had in fact completed his experimental work prior to mailing his letter to *Science*. One can understand that he felt cheated. He had worked for years on purifying vitamin C and had done persistent, careful science. If he had been second to the finish line, it was only by a couple of weeks. He had trained Svirbely, who abruptly ran off to Hungary to join the laboratory of his major competitor. Rubbing more salt in King's wounds, Szent-Gyorgyi not only professed to be uninterested in vitamins but disparaged the entire field.

In addition, while King was toiling away in Pittsburgh, Szent-Gyorgyi came to the answer as much through good luck as good science. This is a common story. Luck is important in science, but a scientist must have the judgment, intelligence, and drive to exploit the good luck if he or she is to garner the praise. Szent-Gyorgyi had them all.

Svirbely in subsequent interviews was not forthcoming about his motives in moving to Hungary nor about the exact chronology of events. Some speculated that he suspected that hexuronic acid was vitamin C but was frustrated by the inability to get pure material in King's laboratory. Therefore he jumped ship and joined Szent-Gyorgyi, who he knew had the pure compound. King no doubt felt that his trainee had stabbed him in the back.

The dispute left enduring scars on all involved. Szent-Gyorgyi never felt comfortable speaking before audiences in the United States, where the popular press had portrayed him as a usurper. King carried a grudge for the rest of his career, and Svirbely never dispelled suspicions of disloyalty to his mentor.

CRYSTALS FROM PEPPERS

Szent-Gyorgyi and King had shown that hexuronic acid was the antiscorbutic vitamin C, but they still did not know its chemical structure. Svirbely's experiments had exhausted Szent-Gyorgyi's meager supply. In Hungary, Szent-Gyorgyi had no access to adrenal glands in large numbers, but as luck would have it—and Szent-Gyorgyi never lacked for luck—Szeged was the center of the Hungarian paprika industry. Szent-Gyorgyi later told the story (perhaps apocryphal) that one evening, when entertaining an especially boring dinner guest, his wife served fresh paprika peppers.[13] Szent-Gyorgyi did not feel like eating the peppers or enduring the dinner-table conversation. He escaped out the back door, peppers in hand, and retreated to his lab to test them for hexuronic acid. He labeled this act "a husband's cowardice."

Whether or not the story is true, paprika proved to be a rich source of vitamin C. Importantly, the vitamin C in ground-up paprika is much more stable than in citrus juice. Szent-Gyorgyi promptly turned his laboratory into a paprika processing assembly line, and he soon crystallized enough hexuronic acid to allow Norman Haworth, his chemist collaborator in Birmingham, to determine its molecular structure.[14] At the suggestion of a colleague, they changed the name from hexuronic acid to ascorbic acid, as it was the antiscorbutic factor.[15]

For this work, Szent-Gyorgyi received the Nobel Prize in Physiology and Medicine in 1937, and Haworth the Nobel Prize in chemistry the same year.[16] This rekindled the dispute over who had been the first to show that hexuronic acid was vitamin C. The local Pittsburgh press was especially incensed at what it saw as a miscarriage of academic justice. The *Pittsburgh Post-Gazette* claimed to have proof that King, not Szent-Gyorgyi, was the first to show that hexuronic acid was vitamin C, and they made Svirbely the villain. According to the newspaper, "not until a student of Dr. King's, Dr. J. L. Svirbely, carried the knowledge gained in King's laboratory across the ocean to Gyorgyi's laboratory did Gyorgyi discover vitamin C and go on to win the 1937 Nobel Prize in medicine." Svirbely denied disclosing secrets to Szent-Gyorgyi.

King was superficially gracious. He said that perhaps Szent-Gyorgyi had deserved the Nobel Prize not just for vitamin C but for the totality of his work on metabolism. However, in the same breath he cited a letter to *Science* by a nutritionist at the Mellon Institute, a nearby Pittsburgh institution, which again laid out a chronology that awarded priority for the discovery of ascorbic acid to King.[17]

To his credit, Szent-Gyorgyi did not seek a patent on his discoveries or his method for preparing ascorbic acid from paprika, which later became the commercial source of the vitamin. He provided ascorbic acid to other scientists and to supplement the nutrition of children who lived in northern latitudes and suffered vitamin deficiency in the winter. He did, however, patent and manufacture a canned spread for one's morning toast made from paprika and rich in vitamin C. He first named it "Vita-prik" but changed the name to "Pritamin" when friends explained why it did not sell well in English-speaking countries.

Szent-Gyorgyi began to proselytize the benefits of the vitamin. He turned from disparaging nutrition research to touting vitamin C as the answer to multiple health problems. With his newfound scientific prominence, he had speaking engagements throughout Europe and used them as opportunities for "preaching vitamin C," as he put it. As a harbinger of things to come, he advocated, with no animal data and without success, for a clinical trial to test ascorbic acid in preventing colds and infections in babies.

Hoping to follow up on his success with vitamin C, Szent-Gyorgyi became interested in antioxidant compounds termed bioflavonoids, which are found in many foods. He dubbed them "vitamin P" and claimed they could cure colds and confer many other benefits. These claims did not pan out, and he eventually turned his attention elsewhere. He went on to a productive career investigating cellular metabolism and the mechanism of muscle contraction. In contrast to his professional success, his personal life remained bumpy; he had a penchant for changing wives as well as jobs.

Charles Glen King moved to Columbia University, where he continued his distinguished career in biochemistry. Joseph Svirbely returned to Pittsburgh for a few years and then engaged in toxicology research for US government laboratories. He kept detailed records of his work on vitamin C, including press clippings, but he never explained his motivation for moving from Pittsburgh to Szeged.

This story is a prime example of the role of serendipity in science. If Szent-Gyorgyi had not wandered into the right laboratories in his peripatetic travels across Europe after World War I, if Svirbely had not walked into his laboratory in Szeged, if Szent-Gyorgyi had not escaped a boring dinner guest to go his laboratory, paprika peppers in hand, and if those peppers had not been a gold mine of vitamin C, it is likely that Szent-Gyorgyi never would have completed the race, much less won. In science, as in sports, sometimes winning is more a matter of luck than skill. But as Szent-Gyorgyi's story also demonstrates, it requires a nimble mind to exploit good fortune.

Once the chemical identity of vitamin C was known and ascorbic acid was produced commercially, scientists could apply the burgeoning armamentarium of biochemical techniques to understand its role in physiology and metabolism. This was normal science, and it illustrates the power of that endeavor. Paradigm shifts are not required to make large strides in understanding nature. Even though no more Nobel Prizes have been awarded for vitamin C, it has been a period of tremendous progress.

SCURVY FOR SCIENCE

Well, it's always the patient who has to take the
chance when an experiment is necessary. And we
can find out nothing without experiment.

—George Bernard Shaw,
The Doctor's Dilemma, 1906

Investigators have deliberately produced scurvy in volunteers. One purpose of these experiments was to study scurvy under controlled conditions, to be able to draw blood, collect urine, and perform tests on the volunteers. Another aim was to determine the minimum amount of vitamin C required to prevent scurvy as a step toward defining a minimum daily requirement for the vitamin.

SELF-EXPERIMENTATION:
AN INAUSPICIOUS BEGINNING

In two cases, the investigators experimented on themselves. That effort got off to a bad start with William Stark.[1] Dr. Stark was born in Birmingham in 1740 and educated in Glasgow and Edinburgh. He moved to London to undertake research with the renowned surgeon and pathologist, John Hunter. He met Hunter's friend, Benjamin Franklin, who was living in England at the time. Among the many subjects that interested Franklin was nutrition. He told Stark that he had once eaten only bread and water for a few weeks and felt fine the whole time. He wondered whether any other food was necessary.

That conversation spurred Stark to tackle this question using himself as the experimental subject. Over a period of more than nine months, he ate a variety of restricted diets and kept a journal describing the effects on his health and mood. He began in June 1769 with Franklin's bread-and-water diet. In the middle of August, after ten weeks on the diet, his gums swelled, and he began to develop signs of scurvy. He never stated in his journal that he recognized his diagnosis.

He had frequent contact with the noted physician Sir John Pringle, a friend of Benjamin Franklin. Pringle wrote about scurvy and was never hesitant to pontificate on the subject when given the opportunity, but it is unlikely that he had any firsthand experience with the disease. Pringle served as president of the Royal Society from 1772 to 1778, a tenure mainly remembered for his habit of falling asleep during meetings.[2]

After developing the first signs of scurvy, Stark varied his diet, adding and subtracting meats and fatty foods, including butter and cheese, but never fruits or vegetables. His symptoms of scurvy waxed and waned until he died on February 23, 1770, at the age of twenty-nine. He may have not died of scurvy, as he was generally malnourished and may have had an intestinal infection, but he certainly died with scurvy. Dr. Pringle seems to have been of no help. William Stark was admirable for his courage, persistence, and dedication to science but not for his good sense.

A BETTER OUTCOME

By the late 1930s, after vitamin C and other vitamins had been identified, scurvy could be studied under more controlled conditions than Dr. Stark endured. Using diets lacking in vitamin C but otherwise nutritionally complete and supplemented with other vitamins, pure vitamin C deficiency could be studied in the absence of other nutritional deficiencies that had often confounded historical cases, including Stark's.

In 1939, John H. Crandon, a surgical resident at the Boston City Hospital, was undeterred by Stark's experience. He followed in the tradition of self-experimentation and placed himself on a diet lacking milk, fruit, or vegetables. He lived mainly on cheese and crackers supplemented with all known vitamins except C.[3] He initially recruited three other young men to participate in the experiment but felt he could not ask them to endure hardships he was not willing to undergo himself and joined them as a subject. However, the other three were spotted in a café near the hospital cheating on their diet and had to be dismissed from the study, leaving Crandon as the sole experimental subject.

Chemical assays for ascorbic acid had been developed, and Crandon's blood levels of the vitamin were measured periodically. His plasma ascorbic acid level fell to zero after forty-one days, and his white blood cell content fell to zero after eighty-two days. The white blood cell levels are a better indication of the amounts of vitamin stored in bodily tissues than the plasma levels.

After four months on the diet and a month after his white blood cell level of ascorbic acid had become undetectable, he began to feel weak and easily fatigued. His endurance on a treadmill was impaired. He tried to continue performing his duties as a surgical resident but began to nap in the afternoon and to ask colleagues to attend to his patients when he was too tired. He gained the reputation of a slacker among his fellow residents.[4]

On day 134 of the diet, almost two months after his white blood cell ascorbic acid had become undetectable, he developed the first physical signs of scurvy. Dark spots appeared around hair follicles on his legs and the hairs split and became ingrown. After 161 days, he developed tiny hemorrhages in the skin, termed *petechiae*, which first appeared on the lower legs after he had been standing for some time and gradually progressed upward. They reached his thighs after six months on the diet.

As a surgeon, Crandon was interested in wound healing. One of the consistent features of historical descriptions of scurvy was the failure of new wounds to heal and the opening of old wounds, in some cases wounds that had healed years previously. Crandon had incisions made on his back after three and again after six months on the vitamin C deficient diet. The wound made at three months healed normally, even though he had undetectable levels of ascorbic acid in the blood.

The second wound made after six months was biopsied ten days later. The skin was healing normally, but the deeper tissues showed no healing. His appendectomy scar from childhood began to open. Immediately after the biopsy, Crandon began to take one thousand milligrams of ascorbic acid daily. When the wound was biopsied after another ten days, it was healing normally.

In contrast to the experience of sailors, Crandon's gums did not swell or bleed, although they may have become slightly soft by the end of the experiment. Dr. Crandon presumably practiced better dental hygiene than sailors in the seventeenth and eighteenth centuries. It would be no surprise that they did not floss regularly. Although he lost twenty-seven pounds, he did not become anemic; he showed no signs or symptoms of decreased immune function nor did routine laboratory tests become abnormal. Within twenty-four hours of his first dose of ascorbic acid, the subjective weakness abated, and the skin changes resolved over three weeks.

MORE DEMANDS OF WAR

Other experiments followed in which experimental subjects were confined under close observation. During World War II, thirty-five conscientious objectors lived for up to four years in a house christened the Sorby Research Institute in Sheffield, England. Among other duties, they served as subjects for medical experiments.[5] During the war, Great Britain suffered a food shortage, and the government supported efforts to define minimum nutritional requirements to guide food rationing for civilians and provisioning for the military. Twenty of the Sheffield residents took part in an experiment to determine the minimum amount of vitamin C necessary to prevent scurvy.

Throughout the experiment, all twenty volunteers shared an identical diet containing less than one milligram of vitamin C per day. To ensure that they began with adequate body stores of the vitamin, they all took seventy milligrams per day of ascorbic acid for six weeks. At that point, still receiving the vitamin C deficient diet, they were divided into three groups: three subjects continued to receive seventy milligrams of supplemental vitamin C daily and served as the control group, seven subjects were reduced to ten milligrams supplemental ascorbic acid per day, and ten subjects received no vitamin C supplementation. To study wound healing, a three-centimeter incision was made in their thighs and biopsied at various times during the experiment.

All the subjects receiving no vitamin C supplementation developed scurvy after four to six months. Their signs of scurvy included petechial hemorrhages in the skin, bleeding gums, and swollen knees. They complained of aching backs, joints, and limbs. The surgical wounds spontaneously opened and bled. None of the subjects in the other two groups receiving vitamin C supplementation developed scurvy. Ten milligrams per day of vitamin C was sufficient to prevent the disease.

To the surprise of the researchers, two subjects with scurvy developed chest pain and shortness of breath accompanied by changes in their electrocardiograms. The investigators thought these symptoms were caused by hemorrhages into the heart muscle or perhaps between the heart and the pericardium, the lining around the heart. Both subjects received large doses of vitamin C and recovered. These unexpected and potentially life-threatening complications took the investigators aback, as they had no intention of risking the lives of the volunteers. When another subject developed chest pain, he immediately received ascorbic acid even though the pain was attributed to an infection. The investigators were not going to take more chances.

No difference was noted in the health of the groups receiving ten and seventy milligrams per day of vitamin C. Some subjects in the group that developed scurvy were given ten milligrams per day of vitamin C and recovered completely, although slowly. The investigators concluded that ten milligrams per day of vitamin C prevented scurvy, with no apparent benefit of higher doses. However, to provide a margin of error, the Medical Research Council recommended a minimum daily requirement of thirty milligrams per day of vitamin C (later increased to forty milligrams per day).

PATIENTS AND PRISONERS VOLUNTEER

Other experiments produced similar findings but with more detailed study of the metabolism of ascorbic acid. In 1944, Michael Pijoan and Eugen Lozner, US Navy physicians, studied six men and women who were hospitalized for other reasons but volunteered to stay in the hospital and undergo metabolic studies of vitamin C.[6]

The six subjects were first given large amounts of vitamin C to fill their bodily stores and then placed on a diet lacking vitamin

C. Ascorbic acid fell to undetectable levels in the plasma after two to four months and in white blood cells after four to six months. Scurvy appeared after five to six months, although only one subject developed abnormalities of the gums. Based on indirect evidence from their metabolic studies, the investigators estimated that twenty milligrams per day of vitamin C would prevent scurvy and eighty to one hundred milligrams per day would fill the body's stores completely.

Beginning in 1966, two additional experiments were performed with prisoners from the Iowa State Penitentiary who volunteered to be confined to a metabolic ward in the hospital rather than their jail cells.[7] The experiments enrolled a total of twelve prisoners, but two took the opportunity to escape, preferring not to be confined at all.

Five of the Iowa subjects underwent a series of behavioral and psychological studies. The most notable finding was the lack of impairment on several tests of mental function and motor dexterity. Tests of physical strength were impaired only in the legs and only after the subjects developed joint pain. The subjects experienced diminished alertness and attention once scurvy developed. A personality test (the Minnesota Multiphasic Personality Inventory, or MMPI) revealed signs of "depression, social introversion, hypochondriasis and hysteria." These all resolved with repletion of their vitamin C.

Although the signs and symptoms of scurvy were relatively consistent in all the experiments with human subjects, the time from the beginning of vitamin C deficient diets to the first appearance of scurvy varied considerably. The Iowa experiment found considerable subject-to-subject variability in how rapidly the body stores were exhausted. Also, the variability is at least partly a result of the different diets the subjects ate prior to beginning the studies. Those who began with marginal stores of the vitamin, including the Iowa prisoners and Dr. Stark, developed scurvy after only two to three months. In contrast, the subjects of Pijoan and Lozner and the Sheffield volunteers who received ample vitamin C prior to being deprived of the vitamin remained healthy for four to six months before scurvy appeared.

The Sheffield experiment demonstrated that ten milligrams vitamin C per day prevents scurvy in otherwise healthy adults and cures it once it develops. This is the amount contained in a couple tablespoons of fresh orange juice, a small, cooked potato, or a small serving of raw spinach (see the appendix). Whether this is all that is necessary to maintain optimal long-term health is a question that continues to spur controversy.

These studies substantiated the careful clinical descriptions of Lind and other early observers and also found that blood levels of vitamin C fall to undetectable levels weeks or months before scurvy develops, indicating that measurements in the blood are not reliable indicators of scurvy. The studies also verified that poor wound healing is a core feature of the disease, but only after skin changes have been present for weeks.

WAS IT WORTH IT?

These experiments were performed at a time when the standards for clinical trials were much less strict than today. None of these studies would meet current standards for ethical review, methodologic rigor, statistical analysis, or reporting of results. Perhaps we should not judge prior generations by current standards. Nevertheless, whether the knowledge gained from these studies justified the risk and discomfort to which the subjects were subjected remains an open question.

In these studies, experimenters deliberately caused a painful, debilitating, and possibly life-threatening disease in normal, healthy young people. This violates a fundamental principle of medical practice: first do no harm. Two physicians experimented on themselves, and the Sheffield study was motivated by a wartime emergency. These circumstances may mitigate ethical reservations. Another mitigating factor was that the twentieth-century investigators had an infallible cure available. When subjects developed alarming symptoms, high doses of ascorbic acid relieved their symptoms within days.

The coronavirus epidemic has made the ethics of this kind of clinical trial a matter of public debate.[8] In the face of a worldwide crisis, should healthy volunteers be deliberately exposed to the virus to speed the development of vaccines and treatments and possibly save thousands of lives? If unvaccinated controls are used, some experimental subjects will develop COVID-19 disease. This is an especially vexing question since, unlike scurvy, a reliable treatment for COVID-19 is not available as of this writing. Subjects could die or suffer long-term damage as a consequence of the experiment.

On the other side of the coin, volunteers in a study could benefit from being the first to receive a successful therapy. In the scurvy experiments, the subjects had nothing to gain from their suffering except the satisfaction of contributing to knowledge and helping their nation during a crisis. Ethicists disagree about whether this type of study is acceptable, but when countries feel threatened, whether by a foreign power or an epidemic disease, the deliberations of ethicists may take a back seat.

Scurvy will probably never again be intentionally produced in humans for scientific purposes. Although experimentation on human subjects will always have a place in medical science, during the past half-century, the major discoveries in vitamin research have come from the laboratory, not the metabolic ward.

8

NORMAL SCIENCE

Ascorbic acid acts at the foundation of life.

—Albert Szent-Gyorgyi,
The Living State and Cancer, 1978

Since the identification of vitamin C as ascorbic acid and its chemical characterization in 1933, scientists have fleshed out its role in biology. This period of "normal science" has lacked drama, but the scientific advances have been substantial. Although questions remain, a host have been answered.

WHAT IS VITAMIN C?

The chemist Norman Haworth, Albert Szent-Gyorgyi's collaborator, won the Nobel Prize in Chemistry in 1937 for finding the structure of ascorbic acid. It is a sugar-like molecule that contains six carbon, eight hydrogen, and six oxygen atoms (chemical formula $C_6H_8O_6$).[1] Four of the carbon atoms plus an oxygen atom form a ring, and the other two carbon atoms are in a side chain (see the following figure).

Ascorbic Acid
(Reduced Form)

Dehydroascorbic Acid
(Oxidized Form)

Chemical structures of the two forms of vitamin C.

Source: H. Abozenadah, A. Bishop, S. Bittner, and P. M. Flatt, "Chemistry and the Environment," CC BY-NC-SA (2018), https://wou.edu/chemistry/courses /online-chemistry-textbooks/ch150-preparatory-chemistry/

Its resemblance to sugars made it difficult to purify from lemon juice, in which sugar molecules are much more abundant. This hurdle allowed Szent-Gyorgyi, who used bovine adrenal glands and Hungarian paprika as his starting materials, to leapfrog his competitors.

When oxygen reacts with ascorbic acid, it removes two hydrogen atoms, forming dehydroascorbic acid. Inside cells, there are enzymes that convert dehydroascorbic acid back to ascorbic acid. Because of this interconversion, dehydroascorbic acid is also antiscorbutic. Strictly, the term vitamin C refers to both compounds.

When ascorbic acid is not inside a cell—for example, in fruit juice or when pure ascorbic acid is in solution—oxygen in the air reacts with ascorbic acid, and the resulting dehydroascorbic acid is not converted back to ascorbic acid. Instead, it is unstable and breaks down into smaller, inactive molecules. This accounts for the loss of antiscorbutic efficacy of citrus juice in the presence of air. That loss of activity accelerates as the temperature is raised. James Lind's method of preparing his rob—heating lemon juice in an open vessel for hours—rendered it useless to treat scurvy.

To store an ascorbic acid solution, it needs to be protected from oxygen. It should be made acidic, placed in a filled and tightly sealed container to minimize exposure to air, protected from light, and kept at a low temperature. Solid ascorbic acid, in the form of pills or powder, is stable if kept dry.

WHY ALL ANIMALS REQUIRE VITAMIN C

Ascorbic acid permits animals to live in an atmosphere that contains 20 percent oxygen.[2] Oxygen is a double-edged sword for land-dwelling creatures. On one hand, they cannot live without it. Cells need oxygen to generate energy to support cellular metabolism, muscle contraction, and neuronal activity. In essence, cells burn carbon-containing foods. However, rather than the energy being dissipated as heat, it is retained to drive cellular processes. When this mechanism evolved, it permitted animals to generate sufficient energy to become mobile and manipulate their environment.

On the other hand, oxygen can be dangerous. Unregulated oxidation degrades cellular proteins, lipids, and nucleic acids.[3] Not only are we exposed to atmospheric oxygen, but normal cellular metabolism produces many oxidizing agents, termed reactive oxygen intermediates (ROI). Molecules essential to cellular function must be protected from these marauding, highly reactive molecules. Among other adverse effects, oxygen may facilitate the deposition of fats in blood vessel walls and cause the degeneration of neurons. Gradual and relentless oxidation may be a component of the aging process.

The major antioxidants in cells are ascorbic acid and glutathione (GSH), a molecule comprised of a chain of three amino acids. There is about ten times more glutathione than ascorbic acid in cells, and it mainly exists in its reduced, antioxidant form.[4]

Glutathione acts as a loyal bodyguard by initially taking the hit and shielding essential molecules from the attack of oxidizing agents. However, ascorbic acid is the eventual fall guy when the oxygen gets passed on from oxidized glutathione. Ascorbate participates in a reaction that returns the oxidized glutathione back to its antioxidant form while ascorbic acid is oxidized to dehydroascorbate. In scurvy, when there is insufficient ascorbic acid to react with the oxidizing agents, those agents may damage tissues and account for some features of the disease.

There is one additional complication. At high concentrations—higher than exist in the body under normal circumstances—ascorbic acid may become prooxidant. Therefore it may facilitate oxidation rather than protect against it.[5] This only comes into play when high doses of ascorbic acid are administered intravenously. With oral ascorbate, not enough gets absorbed from the intestine to achieve prooxidant levels in the blood. This prooxidant action of ascorbic acid may be toxic to cancer cells, and clinical trials are testing high-dose intravenous ascorbate for that purpose.

HOW THE ABILITY, AND THEN THE INABILITY, TO MAKE ASCORBIC ACID EVOLVED

The first animals that lived in the oceans did not synthesize vitamin C, and today few species of marine animals make the substance. However, soon after the first amphibians ventured onto land, they evolved that ability and passed it on their descendants, including mammals.

There are exceptions. Several animal species have lost the ability to make vitamin C.[6] In addition to primates, these include guinea pigs, almost all species of bats, and some species of birds. In these animals, the gene for the final enzyme in the synthetic pathway of ascorbic acid is mutated to an inactive form. This enzyme has the catchy name L-gulano-γ-lactoneoxidase (GULO).

Inactivating mutations in the GULO gene have occurred several times during evolution and have had no deleterious effects if the diet contains the vitamin in abundance. In technical terms, the mutations are neutral. Species that are primarily vegetarian and that have access to dietary sources of abundant vitamin C can prosper despite losing the ability to synthesize the molecule, although they are susceptible to developing scurvy when the diet is deficient.

WHAT SCURVY DOES TO ANIMALS THAT CANNOT SYNTHESIZE VITAMIN C

Axel Holst and Theodor Frølich fortuitously chose guinea pigs as their experimental animals. The guinea pig lacks an active GULO protein and develops signs of scurvy when limited to a diet lacking vitamin C. They first develop swelling of joints and lie on their sides with the affected limb held in the air, keeping their weight off it. Their teeth and gums become affected, and they may press the side of their face into the floor of the cage. They have hemorrhages into their limbs, and the growing ends of the bones separate from the shafts. This last finding was crucial in identifying the condition as scurvy. James Lind and Thomas Barlow described similar abnormalities of the bones in human scurvy.

Reflecting the more rapid metabolism of small animals, the time course of the disease is compressed compared to humans. Guinea pigs develop scurvy after two weeks on a vitamin C deficient diet.

Other mammalian species that normally can synthesize ascorbic acid may undergo spontaneous mutations that inactivate the GULO gene. Two such strains of mutant rats, termed ODS and Sfx, were discovered because of bone abnormalities.[7] Molecular biologists have deleted the GULO gene in mice, and these mice are now commonly used in vitamin C research. Monkeys and other apes also lack an active GULO gene, but the expense and difficulties of caring for monkeys have limited their use as experimental animals.

WHY THERE IS SO MUCH VITAMIN C IN PLANTS

All plants synthesize their own vitamin C, and many—including fruits, berries, and green vegetables—make lots of it (see appendix). One orange contains more than five times the daily amount of vitamin C necessary to prevent scurvy in an adult human.

One of its functions in plants is the same as in animals: to protect the cells from oxygen.[8] Whereas animals burn carbon-containing food to produce energy and generate carbon dioxide as a waste product, plants operate in reverse. They absorb carbon dioxide from the air and use energy from sunlight to split the carbon atom from the two oxygens. The carbon gets incorporated into molecules that make up their tissues, and the oxygen is released into the air. This process is termed photosynthesis—the use of light energy to make molecules.

Within plant cells, photosynthesis takes place in chloroplasts, the intracellular organelles that contain chlorophyll and make plants green. This process generates the oxygen we breathe and, at the same time, other highly reactive oxidizing agents. Ascorbate is synthesized in the mitochondria of plants and transported into the chloroplast, where it protects essential cellular machinery from oxidation. For example, the most abundant oxidizing agent produced during photosynthesis is hydrogen peroxide. Ascorbic acid is a cofactor in an enzyme that converts the hydrogen peroxide to oxygen and water, which pass harmlessly into the atmosphere. Since ascorbic acid is present in high concentrations in the green, photosynthesizing parts of plants, it is abundant in dark green, leafy vegetables such as kale and spinach and in the leaves of trees.

Ascorbic acid is also abundant in the pulp of fruits and berries, tissues that do not engage in photosynthesis. Szent-Gyorgyi became interested in oxidation-reduction chemistry when he investigated why some fruits turn brown after exposure to the air and others keep their natural color. Scientific curiosity about a seemingly trivial question can lead to major discoveries. Szent-Gyorgyi quickly found that the fruits and vegetables that did not turn brown contained high concentrations of a reducing agent, which he eventually identified as ascorbic acid. Although that explains why lemon juice prevents a cut avocado from turning brown, Szent-Gyorgyi never explained why ascorbic acid is present in high concentrations in certain fruits but not others, and to this day that is understood only in general terms.

The fruit of a plant contains its seeds and has evolved to be tasty to animals. The animals eat the seeds along with the fruit and then spread the seeds around their territory, mainly in their feces. The antioxidant ascorbic acid in the fruit preserves its tastiness. Pure ascorbate itself has a tart taste that would be unlikely to make the fruit an inviting snack, but it can protect tastier substances, mainly sugars. Hence, vitamin C helps preserve the flavor and texture of fruits and berries. The meat of the apple, for example, oxidizes and turns brown when bruised or exposed to air. This makes the fruit less tasty, and bruised, damaged, or soft apples are avoided. Even in the supermarket, vitamin C is important in counteracting the adverse effects of oxidation.

WHAT VITAMIN C DOES WITH OXYGEN OTHER THAN FIGHT IT

Besides protecting cellular machinery from the damaging effects of oxygen, ascorbic acid also helps harness the beneficial properties of oxygen by acting as a cofactor for enzymes.[9] Enzymes are proteins that catalyze biochemical reactions, and cofactors are small molecules that bind to the enzyme and allow it to exert its full catalytic activity. Ascorbic acid is a cofactor for many enzymes that contain iron or copper. These metal atoms are easily oxidized, as is seen when iron rusts or copper acquires a green patina. Ascorbic acid acts as a reducing agent to maintain the metal atoms in their active, reduced (unoxidized) state. Many of these metal-containing enzymes catalyze the transfer of oxygen atoms to proteins or other molecules in a highly regulated manner.

The role of ascorbic acid in the synthesis of collagen is of direct relevance to key features of scurvy.[10] Collagen is the most abundant protein in animals, and it is the major molecule holding our bodies together. Hair and nails are mainly composed of collagen, and collagen makes skin, cartilage, and tendons tough. It helps maintain the integrity of blood vessels. Collagen synthesis is impaired in scorbutic animals. Many of the manifestations of scurvy result from defective connective tissues: soft gums and loss of teeth, bone defects, breakdown of wounds, broken blood vessels with resulting hemorrhages, and abnormal hair fibers.

Ascorbate stimulates collagen synthesis in multiple ways. The best understood is that it allows the collagen protein to attain its proper shape. Collagen is a fiber composed of three amino acid chains wound around each other to form a triple helix. For the three collagen protein chains to fold and entwine correctly, hydroxyl groups (an oxygen plus a hydrogen atom) must be attached. This reaction is catalyzed by an iron-containing enzyme that uses ascorbic acid as a cofactor.

Also of possible relevance to the symptoms of scurvy, vitamin C participates in the synthesis of carnitine, a molecule essential for muscle cells—both skeletal and cardiac muscles—to generate the energy required to contract.[11] Muscles depleted of carnitine are weak. Guinea pigs placed on a vitamin C deficient diet have low carnitine levels in multiple organs, including heart and skeletal muscles. The weakness and shortness of breath, which are prominent early symptoms of scurvy, may result from a deficiency of carnitine in muscles.

Another mechanism that uses ascorbic acid is that by which cells respond to low oxygen levels.[12] The discoverers of this mechanism won the Nobel Prize in Physiology or Medicine in 2019. When oxygen is abundant, an enzyme that requires ascorbic acid adds oxygen to a signaling molecule called hypoxia-inducible factor (HIF). When the oxygen is attached, it is a signal that there is plenty of oxygen. The cell does not need HIF and chews it up. This is the normal state of most of the cells in our body.

When oxygen levels are low, the oxygen cannot be added, and the HIF builds up inside the cell. It tells the cell to make more enzymes that can generate energy from glucose without needing oxygen and to use less of the machinery that requires oxygen. Without oxygen, the cell must make do with less energy than is optimal. This is stressful to the cell, and the HIF induces the cell to make proteins that help the cell to survive this stress. In GULO knockout, vitamin C deficient mice, glutathione can substitute for ascorbic acid and the mechanism is preserved.[13]

Vitamin C is necessary to produce many hormones. Szent-Gyorgyi found his reducing substance, which turned out to be ascorbic acid, in high concentrations in the mammalian adrenal gland. The adrenal gland sits on top of the kidney and has two layers: the core, or medulla, manufactures and secretes steroid hormones, including cortisol; the outer layer, or cortex, manufactures and secretes catecholamines, mainly epinephrine. The synthesis of both classes of hormones requires ascorbic acid as a cofactor.[14] Ascorbic acid is stored within the same subcellular vesicles, the chromaffin granules, in which epinephrine is stored and secreted. Epinephrine is easily oxidized, and the ascorbate, besides participating in its synthesis, helps to maintain it in its active form.

Vitamin C is also present in high concentration in the pituitary gland, which sits at the base of the brain and secretes several "master hormones" that control the activity of other hormone-secreting organs, including the adrenal and thyroid glands. Ascorbic acid is required for the finishing step in the synthesis of several pituitary hormones.

The brain has high levels of vitamin C compared to other organs, second only to the adrenal gland. In the brain, just as in the adrenal gland, ascorbic acid is a cofactor for the synthesis of catecholamines and protects them from oxidation. In guinea pigs deprived of vitamin C, the brain and adrenal gland hold on to the vitamin more avidly than other organs, reflecting the importance of the vitamin in their function.[15]

VITAMIN C AND INFECTIONS

Claims that vitamin C boosts immunity to infections have generated interest in the role of vitamin C in the immune system. If the claims are true, one would expect that people with scurvy would have been prone to infections. There is little evidence of that.[16] Victims dying of scurvy frequently had pneumonia terminally, but that is true of any debilitating illness.[17] Wound infections were also common, but they were much more likely to have resulted from a failure of wound healing than from immune deficiency.

Two observations indirectly support a role of vitamin C in immune function. First, ascorbic acid concentrations are high in white blood cells—the blood cells that form the first line of defense against invading microorganisms.[18] The concentration of ascorbate inside of the most abundant type of white blood cell, the neutrophil, is fifteen to thirty times greater than the concentration in blood plasma. The concentration is even higher in another white blood cell, the lymphocyte.

The presumed role of vitamin C in neutrophils is to act as a protective antioxidant. A key mechanism by which neutrophils kill invading bacteria and viruses is by engulfing them, walling them off in an intracellular organelle called a phagosome, and releasing a burst of oxidizing agents into the phagosome to kill the invaders.

High concentrations of ascorbate protect the neutrophil from committing suicide in the process.

The second observation is that plasma and white blood cell concentrations of ascorbic acid fall in response to acute infection. In people with colds and animals injected with viruses, blood and white blood cell levels of the vitamin fall rapidly and return to normal upon recovery from the infection.

These observations have prompted a series of studies in tissue culture to examine the effects of ascorbic acid on white blood cells. Replacement of ascorbic acid has several effects on the behavior of leukocytes, which are first made deficient in the vitamin. However, white blood cells with normal stores of the vitamin do not function better if given an extra dose of the vitamin, nor do high doses of vitamin C protect animals against infection or increase the production of antibodies. So if vitamin C protects against infections, the mechanism is unknown.

THE NEW
BUCCANEERS:
THE BUSINESS
OF VITAMINS

9

THE PASSION OF LINUS PAULING

Just when objectivity matters most, scientists great scientists, perhaps, above all—are apt to draw on their deepest rhetorical and political resources to skew the course of inquiry to favor their own ends.

—Michael Strevens,
The Knowledge Machine, 2020

The name most closely associated with vitamin C, at least among those old enough to remember him, is Linus Pauling. He is the man who popularized megadoses of vitamin C to prevent colds. He was a brilliant scientist and a prominent member of the post–World War II peace movement. He is the only person to win two unshared Nobel Prizes, one in chemistry and one for peace. However, his fame is not due to these accomplishments but for his zealotry in promoting vitamin C as a panacea. He claimed that it would not only prevent colds but also prolong life by preventing cancer and heart disease.

Why did this exceptional scientist and political activist veer from a career studying molecular structures to focus on nutrition decades after it had ceased to be at the forefront of biochemical research? And why did he become a zealot in advocating megadoses of vitamin C with only the flimsiest of scientific evidence to back his claims? The answer to these questions requires understanding his personality and his long intellectual path to vitamin C.

PAULING'S EARLY LIFE

Linus Pauling was born in Portland, Oregon, in 1901, the oldest child of a middle-class family.[1] His father, a pharmacist, died when Linus was nine years old, and Linus spent the rest of his childhood in a financially strapped family with a mentally unstable mother and a free-spirited younger sister. He developed an early interest in science—first in mineralogy and then in chemistry. Eager to leave home, he dropped out of high school at age sixteen and entered the Oregon Agricultural College (now Oregon State University) in Corvallis, where he studied chemical engineering. Chemistry was a burgeoning field with strong financial support from western states because of its importance to the mining industry.

Pauling exhibited intellectual promise from an early age. He was an outstanding student in all subjects except physical education. He also exhibited intellectual arrogance from an early age. In a college chemistry course, a professor said that he and Linus agreed on the answer to a problem, and when two great authorities agree, the answer must be right. Linus asked, "Who's the other one?" For another course, he wrote in a paper, "I have attempted to use words of one syllable to as great an extent as is practicable in order to prevent any mental strain."

Pauling went on to graduate work at the then brand-new California Institute of Technology in Pasadena, marking the beginning of a four-decade-long association with that institution. Although just getting off the ground, the chemistry department had attracted Arthur Amos Noyes, the most prominent chemist in the United States, as chair. Noyes recognized Pauling's promise and assigned him to the laboratory of a young faculty member, Roscoe Dickinson, to learn the new technique of X-ray diffraction. Noyes believed it to be the most important technology available to advance the field of physical chemistry.

X-ray diffraction determines the three-dimensional structure of molecules. A beam of X-rays is shot through a crystal of a chemical compound, and the bending, or diffraction, of the X-rays as they glance off atoms generates a complex pattern of spots, lines, and curves on a photographic plate. In principle, by precisely measuring that pattern and applying the correct mathematics, one can deduce the position of each atom in the crystal structure. That is only in principle. The images can be complex and the mathematics daunting; before computers, the measurements and calculations were tedious and time consuming.

Pauling mastered the technique quickly, and true to his nature, he wrote the draft of his first publication as a new graduate student without including his mentor, Dr. Dickinson, as a coauthor. Noyes made sure that Pauling corrected that omission, and the paper appeared in print with Dickinson as first author and Pauling second.

Pauling earned his PhD from Cal Tech in 1925 based on his solving the crystal structure of several chemicals using X-ray diffraction. Besides his laboratory work, Pauling took courses in mathematics and physics. He especially concentrated on quantum theory, which was fomenting a revolution in atomic physics.

Pursuing this interest and with the strong support of Noyes, Pauling received a Guggenheim Fellowship to spend a year at the Institute of Theoretical Physics in Munich to work in the laboratory of Arnold Sommerfeld, one of the major figures in quantum physics. A who's who of international physicists spent time in Sommerfeld's laboratory. Pauling took the opportunity of being in Europe to tour the laboratories of the major figures in atomic physics and introduce himself to the giants of the field. This was probably the only time in Pauling's life that he felt humbled by other minds.

A MATURE SCIENTIST AT A YOUNG AGE

Pauling returned to Pasadena with a faculty appointment as an assistant professor of theoretical chemistry. He set out to apply the tools of the new physics to understanding the chemical bond. He accomplished this in a series of publications culminating in his 1931 landmark paper, "The Nature of the Chemical Bond," published in the *Journal of the American Chemical Society*.[2] This earned him promotion to full professor only four years after his initial faculty appointment. At the age of thirty, he had achieved national academic prominence. Two years later, he would become the youngest person ever elected to membership in the National Academy of Sciences. In the words of his biographer, "he would rarely doubt himself after 1931." In Pauling's own words, "I might well have become egotistical as a result."

In laying a foundation for his future work, he built on his theoretical work by applying his knowledge of the size of atoms, known crystal structures, and the properties of the chemical bond to develop a series of rules specifying how atoms were likely be arranged in crystals. This allowed him to reverse engineer the calculation of crystal structures from X-ray diffraction patterns. Rather than tediously calculating the crystal structure from the diffraction pattern, he went immediately to what he thought was the right answer, and then he checked it against the data. It was much easier to calculate the diffraction pattern that would be produced by his model crystals than the other way around. If that pattern matched what was on the photographic plate, he had found the answer without having to perform many hours of tedious measurements and calculations.

This problem-solving method—jumping to the answer and checking against the data—characterized Pauling's work over the next two decades. When first trying to work out the structure of a molecule, Pauling began with model building. He exploited his huge knowledge base to construct diagrams and make paper models, twisting them and aligning them to see what fit. Only after he had what he thought was the most likely structure based on his rules and his intuition, did he turn to the X-ray diffraction pattern to see if his model fit the data. This strategy allowed him to outpace his competitors, who attempted to methodically derive molecular structures from the diffraction patterns.

Pauling's success as a scientist can be attributed to several characteristics. He was, first and foremost, brilliant and endowed with a prodigious memory. He was fearless in attacking major scientific problems. His strategy of using his own intuition and knowledge of chemical theory to leap to the correct answer served him well. Using this method, he worked out the structure of many molecules, inorganic and organic, including the structure of amino acids, the building blocks of proteins. These successes gave him confidence in his scientific intuition. His confidence never wavered, even after it led to an embarrassing blunder.

PAULING TACKLES BIOCHEMISTRY

After his success in determining the structure of small molecules, he turned to the study of proteins. A protein is composed of one or more chains of amino acids. There are twenty different amino acids that compose mammalian proteins. Each amino acid has a common core structure, allowing it to link to others to form a chain. But they differ by the side chains attached to that core structure. Proteins are like charm bracelets. The amino acids link into a strand with twenty different side chains hanging off. Each protein has a unique sequence of amino acids; the gene coding for the protein determines their order.

In the 1930s, it was not known whether proteins, especially those floating in solution inside the cell, had any definite structure. The first attempts at X-ray diffraction of crystallized proteins produced uninterpretable blobs. With better sample preparation, investigators began to obtain X-ray diffraction patterns indicating an ordered structure. Using his model-building approach and the general features of the X-ray diffraction patterns of pure proteins, Pauling developed proposed chemical structures. He recognized that what maintained the structure of proteins was a type of weak attraction between the side chains, termed the hydrogen bond. Although each individual attraction was weak, when added up along a chain of more than a hundred amino acids, the sum of the attractions sufficed to maintain an ordered structure of a protein dissolved in water inside the cell.

As a member of the National Academy of Sciences, Pauling had the privilege of publishing in its journal, the *Proceedings of the National Academy of Sciences (PNAS)*. In contrast to other scientific journals during this era, *PNAS* virtually never subjected submissions by members of the National Academy to peer review. Perhaps the editors felt that members of the National Academy had no peers and did not need their work checked by lesser scientists.

Pauling availed himself of this privilege and published his proposed structures in a series of nine papers, seven of which appeared consecutively in the same issue in 1951.[3] Two of his structures, termed the *alpha-helix* and the *beta-pleated sheet*, proved to be core features of the architecture of many proteins, representing fundamental insights into how proteins fold. Pauling received wellearned accolades for these elegant models, both from scientists and from the public. A laudatory article appeared in *Life* magazine, featuring a two-page photograph of his beaming face next to a model of the alpha-helix.

Although two of his structures proved correct, the other structures proposed in his series of papers were wrong. Pauling's model-building methods could lead him down the wrong path. He had developed a practice of publishing conjectures without supporting data, taking credit for his correct guesses, and ignoring the incorrect ones. At this juncture he was not punished for his mistakes, as the alpha-helix and beta-pleated sheet attracted all the attention and the erroneous guesses were ignored.

An important offshoot of Pauling's interest in protein structure was his hypothesis, later proven correct, concerning the molecular basis of sickle cell disease (also called sickle cell anemia). Sickle cell disease is a genetic disorder common in regions of Africa where malaria is endemic. It is now known to be the result of a mutation in the hemoglobin gene. People who inherit one copy of the sickle cell gene have relative protection against malaria, but those who inherit two copies have abnormal red blood cells. When the oxygen level in the blood is low, the red cells, normally disc-shaped, shrink and contort into a sickle shape. As a result, they become rigid and sticky. They clump and get stuck in capillaries, producing agonizing attacks of bone and joint pain, strokes, and heart attacks.

In 1949, Pauling and colleagues showed that the defect in sickle cells results from an abnormality in the hemoglobin protein—the insertion of a wrong amino acid. He titled the paper, published in *Science*, "Sickle Cell Anemia: A Molecular Disease."[4] He had made yet another fundamental discovery: an incorrect amino acid underlies many genetic diseases.

HIS BIG BLUNDER

The intuitive model-building approach that had served Pauling well to this point then failed him dramatically. This wrong guess was not forgotten by his fellow scientists.

He published a paper in 1953 in *PNAS* proposing that DNA was a triple helix, three strands of DNA coiled around each other.[5] At that time, it was known that DNA was the material of which genes were made. Its structure and the mechanism by which the genetic information is coded in the DNA are fundamental to all of biology. A paper from one of the most famous scientists in the world proposing a solution to the most important problem in biological science attracted a great deal of attention. If it were correct, another Nobel Prize would have been in store.

But it was not correct. To Pauling's dismay, a few weeks later, two unknown neophytes, James Watson and Francis Crick, published their double helix structure, substantiated by Rosalind Franklin's X-ray diffraction images and providing an elegant picture of the genetic material. Pauling immediately recognized that he had been wrong and that Watson and Crick's double helix was correct. He had embarrassed himself publicly.

In his rush to be first, he had made fundamental errors. First, he ignored an essential aspect of the chemistry of nucleic acids, their affinity for water. His triple helix structure left no room for the water molecules embedded within DNA. Second, his structure provided no mechanism by which genetic information is encoded within the DNA strands.

Most important, he failed to account for a key feature of DNA. Like proteins, each DNA strand is a chain, but of four nucleic acids rather than twenty amino acids. The four nucleic acids are adenosine, guanosine, thymidine, and cytosine. It recently had been discovered that in each DNA strand the number of adenosine molecules always equals the number of thymidines and the number of guanosines equals the number of cytosines. The triple helix structure provided no explanation for this striking feature.

Watson and Crick's double helix provided explanations for all these features of the DNA molecule. They, and not Pauling, would be invited to Stockholm to receive the Nobel Prize.

Pauling's intuition told him that the triple helix was the correct structure, and he contorted his models and his thinking to force them to fit the data. This time his modus operandi led him down the wrong path. His encyclopedic knowledge of chemistry and his intuition concerning chemical structures had failed him. Some began to think the great man had lost his edge.

ORTHOMOLECULAR MEDICINE

Perhaps chastened by the triple helix debacle, Pauling did little original science after his sickle cell work. He turned his attention to broad questions of public health. He devoted most of his time during the 1950s to informing the world about the dangers of radioactive fallout from atmospheric nuclear bomb testing. He neglected his laboratory and administrative duties at Cal Tech while he traveled, made speeches, gave interviews, and used his charisma and teaching ability to convince people that the atom was not always their friend. He honed his skills as an evangelist and enjoyed the experience. He was still a star. He won his second Nobel Prize in 1962, but for peace, not physiology and medicine.

However, Pauling did not lose contact with basic science completely. His interest in sickle cell disease led him to read about another genetic disease, phenylketonuria (PKU), a rare but devastating condition. Children who inherit two copies of the mutant gene lack an enzyme that metabolizes the amino acid phenylalanine. Without this enzyme, compounds termed phenyl ketones accumulate in the blood and are excreted in high concentrations in the urine, hence the name of the disease. The consequence is profound intellectual disability and an early death. The treatment is a dietary restriction of proteins that contain phenylalanine, thereby lessening the burden of the amino acid that cannot be metabolized. The diet is unappetizing, but it modulates the levels of phenyl ketones and ameliorates the worst consequences of the metabolic defect.

Pauling's reading about PKU in the early 1960s led him to speculate that other mental illnesses—perhaps all mental illnesses—resulted from similar defects in metabolic pathways, and they might be treated with manipulations of the diet. Based on his speculation, Pauling and colleagues received a grant to test the urine of mental patients for abnormal substances. Three years of searching produced no useful results.

This did not deter Pauling. He became aware of the claims, based on scant evidence, of the Canadian psychiatric researcher, Abram Hoffer, that schizophrenia could be treated with high doses of the B vitamin, niacin. Hoffer and his colleague, Humphry Osmond, called their treatment "megavitamin therapy." Pauling extended Hoffer's theory and coined the term "orthomolecular psychiatry." By this he meant "that mental illness is for the most part caused by abnormal reaction rates, as determined by genetic constitution and diet, and by abnormal molecular concentrations of essential substances."

Pauling speculated that dietary manipulations, along with megavitamin therapy, could restore the appropriate chemical balance and cure mental illness. He published his theory in a review article in the journal *Science* in 1968.[6] The paper contained no original data, only conjecture. It offered a new paradigm for understanding mental illness and the prospect of dietary cures for those illnesses. Given what little was known about mental illness at the time, it was a reasonable conjecture. It took Pauling down a new path and turned his attention to nutrition.

PAULING MEETS VITAMIN C

In 1966 Pauling gave a talk in New York that attracted the attention of Irwin Stone, a biochemist from Long Island. Stone had studied chemistry in college and called himself "Dr. Stone" on the basis of a PhD from Donsbach University, a now defunct, nonaccredited correspondence school in Huntington Beach, California. Stone had mainly worked in the brewing industry but had a long-standing interest in vitamin C.

Stone viewed scurvy as a genetic disease, just as PKU is a genetic disease. In contrast to PKU, however, the mutation causing scurvy occurred long ago in evolution and is not rare but renders all apes and humans incapable of synthesizing vitamin C. The missing nutrient must be supplied in the diet. By Stone's calculations, animals, such as rodents, which can synthesize their own vitamin C, make much more than humans when their much smaller body weight is taken into consideration. Therefore, Stone reasoned, if rodents evolved to synthesize the appropriate amount of vitamin C, the optimum amount of vitamin C in the human diet is much greater than the minimum amount necessary to prevent scurvy.

After hearing Pauling's talk, Stone wrote him a letter touting the benefits of large daily doses of ascorbic acid. Based on his personal experience taking three grams of ascorbic acid per day, Stone claimed that high doses of vitamin C both reduced the frequency and shortened the duration of his colds. He promised Pauling another fifty years of life if he followed suit. Pauling was skeptical at first, but Stone's ideas fit within his own thinking that dietary manipulation could prevent and cure disease. So Pauling also started taking three grams of ascorbic acid every day. He soon became convinced that his own colds had grown less frequent and less severe. He continued this regimen for the next two years while he promoted his grand theory of mental illness.

By 1970 Pauling had attracted little interest in orthomolecular psychiatry; it had garnered little experimental support and neuroscience was moving in other directions. In addition, his formulation of the quantum nature of the chemical bond had been superseded by a more convenient model. He had been wrong about the structure of DNA. He had spent most of his time during the past twenty years traveling and giving talks around the world, letting his laboratory atrophy. He was no longer on the cutting edge of chemistry, but his ego required that he remain an important scientist. Therefore, he took up the cause of vitamin C, believing he could exert a major impact on public health.

He began not by doing experiments to test his hypothesis and publishing the results in peer-reviewed journals, but by publishing a book, *Vitamin C and the Common Cold*, in 1970. The book was a best seller. In it, Pauling claimed that taking high doses of vitamin C would lessen both the number and severity of colds.[7]

The primary data he cited were from a study published in 1961 by G. Ritzel, a school physician in Basel, Switzerland.[8] Ritzel studied 279 boys, fifteen to seventeen years old, in two five- to seven-day-long ski camps during one winter. He divided the boys into two groups, one of which received a pill containing one gram of ascorbic acid daily and the other received a placebo. Dr. Ritzel does not specify that the study was randomized, but he did say that it was double-blind, in that neither the boys nor the camp doctors were aware of the treatment assignments. The pills were administered to the hundred-plus boys in each camp "under supervision." Dr. Ritzel stated flatly, "There was no opportunity for the subjects to exchange tablets." He apparently had little respect for the ingenuity of teenage boys.

The boys reported their symptoms daily. The main finding was that only about half as many boys in the group administered vitamin C reported upper respiratory symptoms as those in the placebo-treated group. Pauling took this as proof of the effectiveness of vitamin C in preventing colds. However, the study had several limitations, including the subjective reporting of symptoms and the possibility that the blind was partially broken by some boys tasting their pills. A major limitation of the study is that in a ski camp, upper respiratory symptoms can result from a viral illness or from breathing cold, dry air. There was only one fewer boy in the vitamin C group who developed a fever (eight versus nine), suggesting that there was probably a mixture of causes of the sore throats and runny noses.

After publication of his book, Pauling exploited his proselytizing skills. He gave interviews to virtually anyone who would listen, including *Mademoiselle* and tabloid newspapers. By trumpeting his ideas in the popular press, Linus Pauling single-handedly produced a huge increase in the sales of ascorbic acid such that it became difficult for pharmaceutical companies to satisfy the demand.

His book, articles he subsequently published in *PNAS*,[9] and his media presence motivated five placebo-controlled clinical trials during the ensuing five years. These studies tested the effect of daily supplementation with ascorbic acid on the frequency, duration, and severity of upper respiratory illnesses in otherwise healthy people.[10]

Conducting these trials was challenging. Since the average person has only one or two colds a year and the differences between treatment groups was expected to be small, the studies required large numbers of subjects. The outcome measure, the self-report of symptoms, is subjective. To avoid bias, the trials must be double-blind; that is, neither the subjects nor those running the trial can know to which group the subjects are assigned. This requires that the placebo pill be identical in appearance and taste to the ascorbic acid. The assignment to groups, ascorbic acid versus placebo, must be random, and the investigators must define, up front, exactly what symptoms and what length of time constitute a cold.

Collectively, the trials enrolled more than three thousand subjects. They differed in their methods, but all were double-blind and placebo controlled. They all enrolled healthy volunteers, and all used the report of subjective symptoms as the primary outcome measure. Two studies recruited volunteers from the general population of Toronto, Canada; one studied employees at the National Institutes of Health (NIH); two were conducted in boarding schools. Ascorbic acid, one to three grams per day (except one of the boarding school trials, which administered only two hundred milligrams or five hundred milligrams per day), was administered continuously, not merely when cold symptoms appeared. Some trials added additional higher doses when cold symptoms began.

These studies were consistent in finding that ascorbic acid taken prophylactically had little or no effect in preventing colds nor in decreasing their severity. One reviewer, combining the data from all the randomized, controlled studies available in 1975, estimated that the maximum benefit was to prevent one-tenth of a cold per person per year and to reduce the duration of colds by an average of one-tenth of a day.[11] This modest benefit implied that if the healthy population were to take large daily doses of vitamin C, it would have an insignificant effect on public health.

Studies were also performed of high doses of ascorbic acid used therapeutically; that is, only when cold symptoms developed. These studies were all negative. Two studies assessed the effect of pretreatment with three grams per day of ascorbic acid prior to deliberately inoculating volunteers with cold viruses. Neither study found any effect on the chance of getting a cold, although one found a small reduction in symptom severity.

Based on these results, established medical authorities were skeptical of Pauling's assertions. His claims were based more on conjecture than on data, and he predicted much larger effects than were actually observed in the clinical trials. Standard medical publications, including the *Journal of the American Medical Association (JAMA)*, the *American Journal of Medicine*, the *American Journal of Public Health*, and *The Medical Letter*, published articles reviewing the studies.[12] They found the benefit, if any, of vitamin C in preventing or ameliorating the symptoms of colds to be too small to justify the inconvenience and unknown long-term effects of taking large doses of ascorbic acid daily. They were highly critical of Pauling's claims and his analysis of the data.

Pauling had been a controversial figure when he was educating the public about radioactive fallout, and he did not shrink from arousing controversy again. He took the critical reviews as evidence of a conspiracy by the medical establishment. He accused medical practitioners and pharmaceutical companies of not wanting an effective treatment for viral infections, because that would cut their profits. Physicians depended on office visits for their income and pharmaceutical companies enjoyed the profits from decongestants, cough medicines, and other symptomatic treatments. He conceded that some medical practitioners were not driven by greed; rather, they were merely too busy to look at the evidence themselves and accepted the recommendations of subspecialists too blinded by their preconceptions to accept the paradigm shift he was promoting.

Far from being deterred, Pauling escalated his claims. He revised his book in 1976 with the added assertion that vitamin C could prevent influenza, a much more serious disease than the common cold. The book was consequently retitled *Vitamin C, the Common Cold, and the Flu*.[13] He further claimed that ingesting the optimum amount of vitamin C would increase average life expectancy by twelve to eighteen years by preventing cancer and cardiovascular disease, the major causes of mortality in developed countries.

He boldly asserted, "There is, in fact, evidence that ascorbic acid significantly decreases the incidence of and mortality from so many diseases as to lead us to the conclusion that it has value in controlling essentially all diseases." When these benefits were realized by the general population, it would represent a profound improvement in public health and merit a third Nobel Prize. However, he based his claims on speculation and his unfounded theory of orthomolecular medicine, not on scientific evidence.

In the absence of supportive evidence, he had difficulty getting his theories of vitamin C and cancer published in standard medical journals. Even the *PNAS* began rejecting his papers. The NIH rejected his grant applications because of a lack of preliminary data. To Pauling, these rejections constituted more evidence of a conspiracy against his ideas.

THE TRUE BELIEVER

What made this brilliant scientist go far beyond the evidence to proselytize for a treatment that offers little or no benefit and to make wildly exaggerated claims for the vitamin? Foremost was his ego. He had enjoyed the fruits of two Nobel Prizes: the travel, the awards, the dinners in his honor, the admiring audiences, the access to the press, and all the other associated perks. He was unwilling to merely act the elder statesman; he needed to continue to be part of the action and the center of attention.

There was also a background of institutional melodrama. Because of both his political activity and his neglect of administrative duties, he had left Cal Tech in 1961 after forty-one years at the institution. He first moved to the Center for the Study of Democratic Institutions in Santa Barbara, then to the University of California, San Diego, in 1965, and then to Stanford University in 1969. None of these institutions was willing to give an aging and no longer productive scientist the financial support and the laboratory facilities he felt he merited.

As a result, he left Stanford in 1973 and established his own research institute in Palo Alto. He first named it the Institute of Orthomolecular Medicine but later renamed it the Linus Pauling Institute of Science and Medicine. Unable to win government grant support, the institute was forced to rely on small grants from private foundations and direct fundraising from the public, which was never sufficient to support more than a small staff. If he could reestablish himself as a productive scientist, he could attract stable funding and build the research facility of which he dreamed.

The main driver of Pauling's zealotry was embedded in his way of thinking, a strategy of doing science that had served him well for a half century. He first developed his theory, then checked it against the data. Unless the data proved him wrong, he accepted his theory as the truth. In the case of the triple helix, the evidence proved him wrong almost immediately. But in the case of orthomolecular medicine, although there was precious little evidence to support the theory, there was not enough to prove him definitively wrong. To Pauling, therefore, it was the truth.

Pauling dismissed the studies that failed to support his claims. Clinical trials can never prove that a drug is worthless with absolute certainty, and Pauling could always come up with an objection. When a study showed no benefit of megadose ascorbic acid in treating cancer, Pauling objected that the experimental subjects had received prior radiation and chemotherapy, which weakened their immune systems, hence they could not benefit from immune stimulation by vitamin C.[14] When investigators repeated the study with subjects who had never received radiation or chemotherapy, again showing no clinically significant benefit, Pauling's objection was that the treatment with ascorbate was discontinued when the disease progressed, as is standard practice in oncology.[15] He speculated that this caused a "rebound effect" leading to a growth spurt of the tumors. To test this would have required the investigators to continue treatment with ascorbate and withhold proven effective therapy in the face of progressing cancer, violating accepted medical practice and ethical principles of clinical trials.

This back-and-forth could have gone on forever. Clinical trials cannot answer all questions at once. There is always the possibility that some change in the protocol—a higher dose or a lower dose, intravenous rather than oral administration, using different selection criteria for the patients, etcetera—could produce a positive result. But to practitioners of oncology, enough was enough, and the question was settled. If vitamin C had any benefit in the treatment of cancer, it was not of sufficient magnitude to justify withholding proven effective therapy on the basis of Pauling's orthomolecular theories.

To Pauling, orthomolecular medicine was the truth until it was rigorously disproven, and since the logic of clinical research does not permit ironclad disproof, to him it remained true despite the accumulating evidence to the contrary. Pauling had admitted his mistake concerning the triple helix of DNA almost as soon as he saw Watson and Crick's model. In that case, the competing model was clearly superior, but without a competing and obvious superior overarching theory of disease, Pauling stuck with orthomolecular medicine just as Lind had stuck with his theory of putrefaction.

With Pauling, the science of vitamin C had gone full circle and reverted to the thinking of ancient Greece and the Middle Ages, when the theory of the humors reigned in the absence of any superior theory of disease. Orthomolecular medicine, with its theory of chemical imbalances, was not a lot different in concept than the balance of humors. Fifty years ago, when molecular medicine was in its infancy and the workings of the brain and the immune system were just beginning to be unraveled, there was no superior overarching theory.

Pauling remained true to his beliefs. When his wife was diagnosed with stomach cancer, she underwent surgery but declined radiation therapy and chemotherapy. Instead, she took high-dose ascorbic acid and a dietary regimen. She lived for five years. Linus Pauling himself did the same when he was diagnosed with rectal and prostate cancer in 1991. He died in 1994.

In the end, Pauling's proselytizing for vitamin C left a substantial legacy. It was just not the legacy he intended.

10

VITAMINS, BUSINESS, AND POLITICS

When you mix science and politics, you get politics.

—John M. Barry, *New York Times*,
July 14, 2020

THE BIRTH OF AN INDUSTRY

Linus Pauling did not found a new field of medical science as he had hoped. Orthomolecular medicine never became an important model. However, he helped found an industry. Pauling's credibility as a Nobel Prize–winning scientist and his celebrity status gave a big boost to the vitamin industry. Sales of vitamin C tripled during the year following the publication of *Vitamin C and the Common Cold* in 1970. Within five years, an estimated fifty million Americans were taking vitamin C supplements. And it wasn't just vitamin C sales that were exploding. The entire vitamin and nutritional supplement industry entered a period of exponential growth that has continued until today.[1]

Factories, mainly in China, manufacture more than 150,000 tons of ascorbic acid yearly, worth more than $1 billion. Consumers buy about one-third as a vitamin supplement, and the remainder is used in food processing and preservation and in water purification. Except for the years between 2016 and the COVID-19 pandemic of 2020, when vitamin D occupied first place, vitamin C has consistently been the single largest selling product in the vitamin and mineral market. Despite more and more scientific studies questioning its benefit, pharmacy shelves continue to stock vitamin C preparations, and their sales remain robust.

The popularity of taking vitamins did not begin with Pauling. Vitamins had been in the public eye since Casimer Funk coined the term in 1912. In the 1930s, Albert Szent-Gyorgyi, one of the discoverers of ascorbic acid, promoted vitamin C and marketed his Vita-prik paste before moving on to other scientific interests. Adelle Davis, trained in nutrition and biochemistry, became a best-selling author and media celebrity during the 1950s and 1960s, touting vitamins and "health foods" as the keys to a long life, better mood, and boundless energy.

Depending on how it is defined, the total US market for vitamins and food supplements is $20 to $40 billion annually, and it grows at a steady 3 to 5 percent per year. More than half of adults in the United States take at least one vitamin or nutritional supplement, and more than a third take vitamin C in some form.[2] The demographic groups most likely to take supplements are older, affluent, well-educated, white, and female. Among those sixty-five years and older, more than two-thirds take nutritional supplements, and in this age group their use is growing.[3]

In the United States, for those who do not take vitamin C supplements, the estimated median daily amount of vitamin C in the diet is about one hundred milligrams, sufficient to keep the body stores full and at least ten times more than required to prevent scurvy.

However, the vitamin sellers have convinced millions of consumers that without taking extra vitamin C, one is malnourished.

PR AND POLITICS

The vitamin and supplement industry has achieved rapid growth for many reasons. Increasing affluence, rising concern about health and wellness, and the growth of recreational athletics have all contributed. At the same time, conventional medicine has become technology driven and less humane, generating distrust. As long ago as 1906, George Bernard Shaw captured this sentiment in *The Doctor's Dilemma*, writing: "All professions are conspiracies against the laity." Traditional medicine puts patients through unpleasant screening procedures like mammograms and colonoscopy, whereas the supplement industry promises to make you feel better immediately with no poking, prodding, or testing. The high cost of prescription medications, generating huge profits for pharmaceutical companies, has also fueled antagonism toward the medical establishment.

The supplement industry has capitalized on this distrust and has been especially effective in using print advertising, television, and the internet in promoting its products. Celebrities have learned that they can profit from their fame and beauty by selling vitamins and supplements. Movie stars promise that you too can develop a perfect body, have more energy, enjoy better sex, and live longer by using their products.

My friend Bill, the marathoner we met in the introduction, is representative of many consumers. He took vitamin C pills not knowing if they would help or even if any of the vitamin would be retained in his body. The cost was modest and he did not think it would do any harm, but one could make the same argument about virtually any food substance. Why not take supplemental ginger, for example? It is an ingredient in many traditional medicines in Asia and some believe it diminishes the joint pains of arthritis. Its safety with long-term daily use is unknown, but as a natural substance, maybe it is safe. It is worth remembering that hemlock is also natural.

This type of reasoning underpins the industry. It sells products not by providing evidence that they are beneficial or safe but by making broad, unsubstantiated claims: "supports the immune system," "boosts energy," "improves athletic performance," "promotes a healthy liver," etcetera. Since they are natural, they are assumed to be safe. And individual testimonials, frequently by celebrities, take the place of scientific data. Consumers are always looking for the perfect medicine that cures or prevents diseases at a modest cost and with no side effects or risk. The vitamin and supplement industry offers the hope of such medicines.

A key factor in the growth of supplement use is that the industry has deftly avoided government regulation. Some vitamins, including vitamins A, D, and B6 (pyridoxine), are acutely toxic in high doses. Reacting to cases of vitamin A and D toxicity, in 1972 the Food and Drug Administration (FDA) proposed new rules that would require any product that contained more than 150 percent of the recommended daily allowance of any vitamin in a single dosage form to be regulated as a drug. Even though consumers still could take as many doses as they wished, the vitamin distributors saw this as a threat to their sales.

Represented by its well-funded lobbying and public relations arm, the National Health Federation, the vitamin distributors conducted an intense campaign to pass legislation to rescind the rules. Mistrust of doctors and freedom of choice were the themes. The rallying cries were "don't let the Feds take away our vitamins," and "Americans have the right to make their own decisions about their health." TV ads implied that the rule would criminalize the use of vitamins.

The lobbying effort succeeded. In 1972, Congress passed a bill sponsored by Senator William Proxmire of Wisconsin prohibiting the FDA from imposing any limit on vitamin doses. Although human beings evolved to ingest a few milligrams of vitamin C at a time and data concerning the safety of high doses is limited, pills with as much as one gram of ascorbic acid are readily available. Single-dose packs of ascorbic acid powder contain as much as two grams.

This legislative victory was only a hint of what was to come. The greatest legislative triumph of the supplement industry paradoxically stemmed from two disasters.

The first involved L-tryptophan, an essential amino acid. Like vitamin C, humans cannot synthesize tryptophan from other nutrients and it must be obtained from food. The amino acid has two main roles in the body. It is required to make proteins, and it is a precursor of the neurotransmitter serotonin, which is involved in many brain processes, including sleep, pain, and mood regulation. The promise was that by taking extra tryptophan, your brain will make more serotonin, your mood will improve, and you will sleep better. In the late 1980s, tryptophan became a popular sleep aid. Since we all ingest the amino acid in our food, it was assumed to be safe, even in high doses.

However, in 1989 there was an epidemic of a new disease that came to be called the L-tryptophan eosinophilia myalgia syndrome (L-tryptophan EMS). The condition occurred only in people taking tryptophan supplements and was characterized by myalgia (muscle aches), eosinophilia (marked increase in the number of circulating eosinophils, one type of white blood cell), and degeneration of the peripheral nerves. About fifteen hundred people taking tryptophan suffered from the disease and at least thirty-six died; many others were left confined to a wheelchair by nerve damage.

Investigators found that L-tryptophan EMS occurred almost exclusively in people taking tryptophan produced by one factory in China. The factory was producing tryptophan contaminated with a then-unknown substance. It is now known that the contaminant was a compound formed by tryptophan bound to a lipid. The factory in China extracted the tryptophan from bacteria and a lipid from the cell wall of the bacteria attached to the tryptophan. After the FDA removed tryptophan from the market, new cases have been rare.

The second deadly disaster involved a Chinese herb, ma huang, also known as ephedra. The ephedra plant contains the stimulant drug ephedrine as well as six related stimulant compounds termed ephedrine alkaloids. Synthetic ephedrine had been sold as an over-the-counter nasal decongestant, but it caused cardiac palpitations and anxiety so was removed from the market. It was replaced by the less potent drug pseudoephedrine, which is still marketed as a component of cold remedies.

The FDA had banned the combination of pure ephedrine and caffeine because of abuse potential, but since ephedra was an herb, it remained legal even though its main active ingredient was ephedrine. Ephedra in combination with caffeine was sold as a weight loss pill and as an athletic performance enhancer. It became popular after amphetamines had lost favor because of concerns about addiction and abuse.

Ma huang had been used for centuries in China as a nasal decongestant, among other uses, so how could it be unsafe? Consumers assumed that ephedrine in the form of an herbal preparation would be safer than the chemically synthesized drug. However, beginning in Texas, reports started to flood health agencies about serious, sometimes fatal, side effects associated with ephedra. These included irregular heart rhythms, heart attacks, strokes, seizures, and sudden death. Eventually, more than nineteen thousand adverse events, including 164 deaths, were linked to ephedra.

These public health disasters led the FDA to propose legislation to regulate the sales and labeling of nutritional supplements, in particular, requiring the companies selling the supplements to provide evidence of safety prior to marketing a new product. The industry again took up the gauntlet. The National Nutritional Foods Association (NNFA) was the vehicle to mount intense public relations and lobbying efforts. The arguments echoed those that had been used to pass the Proxmire bill: Americans have a right to make their own decisions about health; doctors know nothing about nutrition and cannot be trusted; big pharma does not want nutritional supplements on the market because they will eat into their profits; since supplements are "natural," they must be safe.

In 1994, Congress passed and President Clinton signed into law the Dietary Supplement and Health Education Act (DSHEA). Senator Orrin Hatch of Utah championed the bill. Whether coincidentally or not, Utah is a center of the nutritional supplement industry and Senator Hatch and his family had close financial ties with that industry.

For FDA regulation, the DSHEA established a third category of product, dietary supplements, in addition to drugs and conventional foods. The act defined a dietary supplement as "a product (other than tobacco) intended to supplement the diet that contains a vitamin, mineral, herb or other botanical, or an amino acid." Ignoring the tryptophan and ephedra disasters, the act all but exempted nutritional supplements, including vitamins, from regulation by the FDA.

As one consequence of the DSHEA and subsequent court cases, the FDA suspended its efforts to ban the sale of ephedra. But in 2003, Steve Bechler, a twenty-three-year-old pitcher for the Baltimore Orioles, arrived at spring training overweight and taking ephedra to help get in shape. He died suddenly of heat stroke after a hard workout, and the medical examiner implicated ephedra in his death. The resulting publicity led the FDA finally to ban ephedra sales. It remains the only supplement facing an outright ban by the FDA.

Synthetic ephedrine remained popular, but as raw material for the manufacture of methamphetamine, an illicit drug with much higher profit margins than over-the-counter cold remedies. Tryptophan subsequently returned to the market, but additional cases of L-tryptophan EMS have been rare.

There was a minor tightening of the regulations in 2006 with the Dietary Supplement and Nonprescription Drug Consumer Protection Act. This bill required distributors of nutritional supplements to report serious side effects to the FDA. Additionally, the packaging of nutritional supplements was required to contain the same nutritional information as breakfast cereal and other processed foods.

A POSTSCIENTIFIC WORLD

Because of these acts of Congress, the only regulatory oversight of the safety of vitamins and other nutritional supplements occurs after the fact, when enough victims have suffered harm that the FDA can prove "a significant or unreasonable risk" to public health. In other words, enough people must suffer harm from a product that the FDA is motivated to take a product off the market and fight through the subsequent court battles that the supplement industry is sure to mount.

The current policy is summarized on the FDA website:[4]

FDA is not authorized to review dietary supplement products for safety and effectiveness before they are marketed.

The manufacturers and distributors of dietary supplements are responsible for making sure their products are safe *before* they go to market.

If the dietary supplement contains a *new* ingredient, manufacturers must notify FDA about that ingredient prior to marketing.

However, the notification will only be reviewed by the FDA (not approved) and only for safety, not effectiveness.

Manufacturers are required to produce dietary supplements in a quality manner and ensure that they do not contain contaminants or impurities and are accurately labeled according to current Good Manufacturing Practice (cGMP) and labeling regulations.

If a serious problem associated with a dietary supplement occurs, manufacturers must report it to FDA as an adverse event. FDA can take dietary supplements off the market if they are found to be unsafe or if the claims on the products are false and misleading.

The key provision is "The manufacturers and distributors of dietary supplements are responsible for making sure their products are safe." It is not the FDA's responsibility. The consumer must trust the companies that make and sell the product. To remove a product from the market, the FDA must first establish that the product is adulterated (unsafe) or that the labeling is false or misleading. However, false labeling has an extremely narrow definition and, as long as no claims are made regarding specific diseases, unfounded claims of broad benefit are permitted.

In the case of vitamin C, promotional materials from the distributors typically make broad claims but with a few exceptions avoid claiming that ascorbic acid will prevent or treat any disease, including the common cold. A typical blurb is:

> *Amazing health benefits*: vitamin C (ascorbic acid) plays an important role in the health and function of the immune cells. It promotes immune system health, promotes collagen production, has powerful antioxidant properties, supports healthy brain function and cognition, supports cardiovascular and heart health, supports blood circulation, helps improve mood, supports bone density, helps decrease muscle soreness, and helps increase absorption of calcium and iron.

Some promotional material has specific disclaimers that the claims have not been evaluated by the FDA.

Since vitamins and supplements are treated as food for regulatory purposes, the blurbs refer to servings, not doses. Several distributors prominently state that there are no Chinese ingredients in their product, and some say that their ascorbic acid is purified from plants, not chemically synthesized. Frequently, the products are described as non-GMO and suitable for vegetarians and vegans.

One of the primary intended consequences of the DSHEA is that, for practical purposes, definitive studies of the safety and effectiveness of vitamins, minerals, and other supplements will never be performed. For prescription drugs, the requirement to seek FDA approval prior to marketing a product forces pharmaceutical companies to fund and conduct the relevant studies. There is no such requirement for vitamins.

Partly to address this issue, in 1991 Congress established the Office of Unconventional Medicine (now called the National Center for Complementary and Integrative Health) within the National Institutes of Health to study nutritional supplements and other forms of "alternative medicine." Unfortunately, the center has not made an appreciable impact. It has been underfunded and has frequently squandered its limited funds on poor science.

We have little information about the long-term safety of vitamins and supplements. The packaging is required to contain contact information for the FDA so that consumers can report side effects, termed *adverse events* in pharmaspeak, and the sellers are required to report any serious side effects that come to their attention. The problems with the system are double-edged. On one hand, there is underreporting. The FDA estimates that it learns of only 1 percent of side effects from nutritionals, and most reports provide insufficient information to be helpful. On the other hand, just because an adverse event occurs while a person is taking a vitamin does not prove that the vitamin caused it. If you develop a rash while taking vitamin C, it is not likely that the vitamin caused it, but it also is not impossible because vitamin tablets contain ingredients other than the pure vitamin.

To thoroughly investigate every report of an adverse event would be impractical, so the FDA concentrates on "serious adverse events"—those that lead to hospitalization, require medical treatment, or cause disability or death—and looks for patterns in adverse event reports, such as an unusually high frequency of rashes. This is an imperfect system, but since the sellers of vitamins and supplements do not conduct controlled clinical trials, it is the best the FDA can do.

From 2008 through 2011, there were 6,307 adverse event reports to the FDA for problems associated with vitamins and supplements. The majority (71 percent) were mandatory serious adverse events reports from the industry, and the FDA judged 64 percent of the remainder to be serious as well. Outcomes defining them as serious included hospitalization (29 percent) and life-threatening conditions or death (10 percent).[5] These reports do not prove that every reported side effect was caused by the supplement, but the reports give an indication of the magnitude of the problem. A survey of emergency room records from 2007 through 2015 estimated that twenty-three thousand emergency room visits per year are related to side effects of nutritional supplements.[6]

Besides the lack of safety data, the consumer is relying on the word of the manufacturer that the ingredients are those stated on the label, that the product is not contaminated or adulterated, and that good manufacturing practices have been employed. The consumer who takes a pill labeled as a vitamin cannot know what is contained in the pill or where it came from. From 2007 through 2016, the FDA found 774 products sold as nutritional supplements—mainly for weight loss or to improve sexual or athletic performance—adulterated with prescription drugs.[7] In the case of vitamin C, the parent ascorbic acid is generally sourced from

Chinese factories, which are all but free of any independent oversight. Information has recently come to light concerning the limited oversight of factories in Asia making generic prescription drugs, and the manufacturing of nutritionals is even less scrutinized.[8]

In sum, the vitamin and supplement industry is the Wild West of the health care industry. A pure drug may be banned as unsafe, but the same chemical compound contained in an herbal supplement may be perfectly legal. The onus is on the buyer to assess efficacy and safety of the products. Sellers are shielded from having to provide all but the most basic nutritional information concerning their products. Since most of the ingredients are manufactured in China and India in factories with little regulatory oversight, consumers must proceed at their own risk.

Unfortunately, this is Linus Pauling's legacy. We associate his name with megadose vitamin C, not with his brilliant work in chemistry or in achieving a nuclear test ban treaty. His two Nobel Prizes are all but forgotten. Only *Vitamin C and the Common Cold* is remembered. Students of chemistry no longer read his excellent book, *The Nature of the Chemical Bond*, but many start taking vitamin C pills at the first signs of a runny nose or sore throat.[9]

The story of the vitamin and supplement industry poses a cautionary tale to those who equate scientific knowledge with progress. The leaders of the Royal Navy who ignored the work of James Lind in the latter half of the eighteenth century were mainly victims of ignorance, contradictory advice, and outmoded thinking. In the twentieth century, the DSHEA enacted a policy of deliberately ignoring scientific evidence. The policy was championed by some heroes of the liberal establishment: William Proxmire, Hubert Humphrey, and George McGovern. President William Clinton signed the legislation. It became law more than two decades before Donald Trump took office. Trump and his allies are not the first to incorporate deliberate stupidity into national policy. It is one of the few vestiges of bipartisanship in the political life of the United States.

LESSONS LEARNED

It seems, to all the world, that there is something about the nature of science itself that the human race finds hard to take on board.

—Michael Strevens,
The Knowledge Machine, 2020

IGNORING THE OBVIOUS

The first takeaway from the history of vitamin C is that our pre-conceived notions of reality—our paradigms, in the terminology of Thomas Kuhn—constrain our thinking and prevent us from interpreting evidence objectively. Presented with new information, we try to force it into our mental models of the world, no matter how much bending and twisting we must do. What does not fit, we often ignore. We expend a lot of effort to save our time-honored models before trying to invent or accept new ones. And in the process, we may fail to understand what is right in front of us.

The four-hundred-year delay in understanding scurvy is a prime example. The Nutrition Board of the Medical Research Council of Great Britain in 1936 explained why it took so long to discover vitamins: "The evidence from disease would have led sooner to a conception of these food constituents and their functions but for a not unnatural bias in thought. It was difficult to implant the idea of disease due to deficiency."[1]

This "not unnatural bias in thought" prevented physicians and bureaucrats from seeing what in retrospect is obvious: scurvy is cured by specific foods and is caused by the lack of an essential nutrient. They could not escape the belief that disease had to be caused by an external agent, such as germs, miasma, or a toxin. If the experience of Vasco da Gama had been seen with an open mind, millions of lives could have been saved. It took Christiaan Eijkman a decade to let go of the germ theory and accept a nutritional cause of beriberi. Moreover, Linus Pauling clung to his belief in megadose vitamin C in the face of continuously mounting evidence that he was wrong.

Nevertheless, rare thinkers can avoid these mental constraints and examine the evidence without bias. Gilbert Blane was one of those. He is the hero of the story of scurvy, and I would argue that he is the hero of the entire saga of vitamin C. He ignored theories and only looked at the data. Others—notably Nikolai Lunin and Gerhardt Grijns—also broke away from the established paradigms and pioneered new approaches to nutrition.

One thread that runs through the history of vitamin C is that many thinkers who make the major advances capitalize on chance events. When unexpected events occur, our initial instinct is to dismiss them as flukes and continue down our well-trodden paths of thought. However, some sit up and take notice of those events as essential clues. Blane studied the effects of the vicissitudes of war on the patterns of disease; Grijns recognized the importance of the unintentional change in the diet of his predecessor's chickens;

and Szent-Gyorgyi began grinding up bushels of paprika peppers as soon as, almost by accident, he found them to be a rich source of vitamin C.

What gave these innovators their ability to ignore the prevailing wisdom and stick to the evidence in front of them? Others like them in the history of science—Charles Darwin, Albert Einstein, and Richard Feynman—are those giants in their fields who advanced science in leaps rather than small steps. No one knows what separates these geniuses from the rest of us. It is not merely a high IQ. There are plenty of brilliant people doing scientific research, but only a few make great leaps of understanding.

Except for the rare geniuses, we all suffer from biases in thought. Our brains require a conceptual framework for our perceptions and beliefs; those that do not fit are discarded. This limitation is evident in politics, interpersonal relations, and economics, just as it is in science.

We can only wonder what biases of thought we harbor today that will baffle observers a century from now. We can be sure that there are many. Perhaps the keys to mental illnesses, such as schizophrenia and autism, are lying right under our noses, and we cannot see them because the reigning scientific paradigms place blinders on us. If only we could understand what permitted people like Gilbert Blane to discard these mental blinders and view the data dispassionately, we too might make leaps of progress.

IGNORING THE SCIENCE

Another lesson is that even when science reveals the answer, we may refuse to accept it. Scientists spend their careers searching for truth in the belief that the truth will free us from superstition and irrational assumptions. Experience is otherwise. People and entire nations refuse to pay attention to the evidence. How many of us persist in eating unhealthy foods, fail to wear seat belts in the car, or find reasons not to wear a mask during a respiratory virus epidemic?

Providing scientific evidence of the safety of vaccines and genetically modified foods does not change people's minds.[2] Even worse, when given the data about the safety of vaccines, anti-vaccine parents grow less likely to have their children immunized.[3] Attitudes about climate change have more to do with political beliefs than science. Many supplement users admit that they ignore the scientific evidence that they are ineffective.[4]

This is human nature. We do not change our behavior or beliefs easily. Despite extensive psychological research, no one has a strategy for persuading people to act more rationally.

The disheartening lesson demonstrates how little science matters to the behavior of individuals or nations. The truth does not necessarily set us free. People are bound by politics, religion, and habit more than by science. Consumers waste their money and, in some cases, risk their health by ingesting megadoses of vitamins in a vain effort to treat common diseases and avoid the ravages of aging.

Additionally, Congress passed laws excluding science from government policy concerning vitamins and supplements. Science frequently presents inconvenient truths. When science threatens profits and cherished beliefs, it is ignored. That ignorance may be enacted into law.

SCIENCE MARCHES ON

Nevertheless, a hopeful lesson emerges from this history: medical science can surmount these limitations, although it may take years, even centuries. As the struggles against cancer, HIV, and other chronic diseases have shown, science can be a frustratingly slow path to discovery. Nevertheless, once the barriers to understanding are overcome, the discoveries can have tremendous impact.

At the turn of the twentieth century, vitamin deficiency diseases were still common. Scurvy was not as widespread as it once had been, but it continued to afflict soldiers in World War I. Beriberi remained endemic in Asia. Rickets stunted the growth and deformed the bones of children among the urban poor in North America and Europe. During the first half of the twentieth century, nutrition science eliminated these diseases that killed and crippled millions.

The story of pellagra recapitulates many of the themes in the history of vitamin C. Pellagra is a potentially fatal disease that produces rash, diarrhea, and mental changes. It became common in the southern states of the United States after the turn of the twentieth century when, for millions of Southerners, machine-milled corn meal from factories in the Midwest replaced hand-milled meal produced locally. The milling machines removed the germ from the corn kernels. Removing enzymes and lipids in the germ rendered the meal more stable in storage and easier to ship, but it also removed most of the niacin (vitamin B3). For those who ate a diet of cornbread, fatback, biscuits, gravy, and molasses—mainly the poor and institutionalized populations—there was no other source of niacin. The epidemic of pellagra ensued. Entire hospitals specialized in treating the disease.

Initially, physicians assumed the disease was infectious, echoing the story of beriberi. Evidence rapidly mounted against that idea and in favor of a nutritional cause, but there was fierce political opposition to the idea that pellagra was a form of malnutrition. Southerners were insulted at the suggestion. Politicians and others clung to the fiction that pellagra was infectious, ignoring the evidence. Consequently, nothing was done to change the diets of those at risk, and between 1910 and 1940 pellagra killed one hundred thousand people in the United States.[5] Finally, in the 1950s, when the evidence could no longer be ignored, laws were passed requiring that flour and corn meal be fortified with B vitamins. Although the delay was unconscionable, once the science was accepted, pellagra disappeared from the United States.

Scurvy vanished from the developed world when its cause was understood, and synthetic ascorbic acid became available. Ascorbic acid is cheap to manufacture. It is added to infant formulas and breakfast cereals and frequently used as a preservative for soft drinks. Improvements in transportation and farming practices made fruits and green vegetables available throughout much of the year and over much of the world. So we all ingest much more ascorbic acid than the few milligrams per day necessary to prevent scurvy. With the fortification of milk with vitamin D and flour with B vitamins, rickets and beriberi have vanished as well. Probably only the building of sanitary sewers, the availability of clean water, and widespread immunizations have saved more lives.

We may strenuously resist changes in our thinking and in our behavior, but eventually most of us accept the truth. The five-hundred-year-old story of vitamin C, in this important sense, has a happy ending.

A GUIDE FOR
THE PERPLEXED

*[V]itamins are characterized by the disproportion
between the great importance of their nutritional
functions and the very small amounts necessary
for the adequate fulfillment of those functions.*

—Medical Research Council Special Report
Series No. 167, 1932

What is the bottom line? Should anyone take supplemental vitamin C, and if so, how much and how often?

The average diet in the United States and Western Europe provides about one hundred milligrams of vitamin C per day.[1] The prevention of scurvy requires much less than that—no more than ten milligrams per day for an adult. Because vitamin C is not only contained in fruits and vegetables but is added to many foods and drinks as a preservative, one would have to make a concerted effort to eat a diet containing less than ten milligrams per day. As a result, scurvy has all but disappeared from the developed world. It occurs only in cases of severe malnutrition, such as the most dedicated alcoholics who derive virtually all their calories from drink. Infantile scurvy, or Barlow's disease, has also disappeared, as human breast milk contains adequate amounts of vitamin C and infant formulas are now supplemented with the vitamin.

However, many people like my friends Mary and Bill believe that more is better and that taking far more vitamin C than is necessary to prevent scurvy makes for better health. The main basis for this belief is the hope that supplemental antioxidants may improve resistance to infection or help retard chronic diseases, including heart and lung disease, cancer, cataracts, and macular degeneration of the retina. The ability of vitamin C alone to prevent these diseases has not been well tested, but clinical trials of combinations of antioxidants, some of which have included vitamin C, have been carried out. These trials are difficult and expensive. Only one or two doses can be tested in any one trial, and the trials face the challenge of controlling the subjects' dietary intake of vitamins. It is no surprise that we do not have an answer for how much vitamin C is optimal.

Given that definitive data are not available and will not be forthcoming soon, does it make sense to take supplemental vitamin C? And if so, how much and on what schedule? The answer to the first question is not straightforward. The second is easier.

WHAT HAPPENS TO VITAMIN C AFTER YOU SWALLOW IT

Our intestines evolved to absorb vitamin C from food. To accomplish this efficiently there is a transport mechanism in the intestinal lining that carries vitamin C across the intestine and into the blood.[2] However, food provides at most tens of milligrams of vitamin C at one meal and the transport mechanism can carry only these small quantities. In technical terms, it is saturable, meaning that there is a maximum amount it can carry into the circulation. Before the

1930s, when factories started producing ascorbic acid, no human intestine had ever encountered hundreds or thousands of milligrams of vitamin C at one time, so there was no opportunity for a mechanism to evolve to deal with megadoses.

When the intestinal transport mechanism is saturated—meaning overloaded—the unabsorbed vitamin C remains in the intestine. Some can bypass the transport mechanism and diffuse across the intestinal lining into the bloodstream, but diffusion is inefficient. What does not make it into circulation continues through the intestine unabsorbed. To decide how much vitamin C to take, we need to know exactly how much vitamin C the transport mechanism can carry and what happens to the rest.

The branch of pharmacology that answers these questions is known as pharmacokinetics, the study of the absorption, metabolism, and excretion of drugs. The information gained can then be used to determine the best dose and schedule of administration. The same approach can be applied to vitamins. Mark Levine and colleagues at the National Institutes of Health (NIH) performed detailed pharmacokinetic studies of vitamin C, published in 1996 and 2001.[3] These studies form the basis for the recommended daily allowance in the United States.[4]

Levine and colleagues studied seven men and fifteen women, all volunteers in good health and between nineteen and twenty-seven years of age. They lived in a hospital metabolic ward for up to nine months and all ate a basal diet containing only five milligrams of ascorbic acid per day. Unlike some prior studies, the subjects were not made vitamin C deficient, as the basal diet was supplemented with varying doses of oral ascorbic acid administered twice a day. To determine the fate of the vitamin, the ascorbic acid content of their blood and urine was measured.

In these studies, a single dose of two hundred milligrams was completely absorbed, but higher doses were only partly absorbed, so that less than half of a 1,250 milligram dose entered the blood.

The unabsorbed vitamin continues through the intestine and eventually down the toilet.

For the vitamin that is absorbed into circulation, other transport mechanisms carry the vitamin from the blood into cells to do its job. The amount in white blood cells is a good measure of the total body stores of the vitamin. It reaches a maximum at a dose of one hundred milligrams per day. Once the stores of all the body's organs are filled, no more can be taken up into cells, and the excess vitamin in the blood is excreted in the urine.

Studies with tracer amounts of radioactive ascorbic acid give researchers another method to follow the fate of a drug or vitamin given to an animal or a person. The amount of radioactivity is typically less than one would receive from a chest X-ray, but there are sensitive methods to detect this radioactivity. These studies verify that about one hundred milligrams per day of vitamin C provides all the vitamin C the body can hold.[5]

Hence, the amount of vitamin C in the average American diet, one hundred milligrams per day, keeps one's cell stores completely filled. Unfortunately, my friends Mary and Bill, both of whom eat a diet containing more than one hundred milligrams of vitamin C per day, have been excreting all their vitamin C supplement into their toilets. To believe that taking additional vitamin C is beneficial is to assume that vitamin C produces some effect that does not require it to remain in the body, perhaps an effect within the intestine or some benefit of simply passing through the blood on its way to the urine.

Although there is no hard evidence of such benefits, there are mechanisms by which vitamin C passing transiently through the intestine and circulation, but not remaining in the body, could possibly exert an effect. Perhaps the spikes in plasma concentration could cause vitamin C to diffuse into cells and exert some temporary effect even though the vitamin will not be retained. Or perhaps megadoses of vitamin C can exert an effect within the intestine without being absorbed. Currently, there is a great deal of interest in the microbiome, the population of bacteria that lives in our colons and affects our bodies, especially in regard to immune function. Vitamin C sloshing through the colon could modify the microbiome in some yet unknown way.

These speculative effects, even if they occur, are not necessarily beneficial. They may also be harmful. Without hard evidence one way or the other, an optimistic assumption is that good and bad are about equally likely.

THE RECOMMENDED DAILY ALLOWANCE

Based on the kind of data collected by Levine, nutritional advisory bodies of various countries have made assessments of a recommended daily allowance (RDA), the amount of a vitamin or mineral that should maintain good health without causing side effects. The recommendations are somewhat arbitrary and vary from country to country. In the absence of definitive studies demonstrating the optimum amount of vitamin C for good health, health authorities must make some assumptions to issue their recommendations.

The involvement of governments in making recommendations about nutrition began at the turn of the twentieth century with the quantification of "total nutritional requirements." There was also an understanding that a varied diet, particularly a variety of proteins, is required to maintain health. The discovery of vitamins led to studies of minimum requirements to fend off vitamin deficiency disease. The National Research Council of Great Britain began to make nutritional recommendations in the 1920s. The involvement of government increased in Britain during World War II, when food shortages motivated the government to both fund research and make recommendations to both military and civilian personnel concerning the minimum nutritional requirements. After World

War II, governmental health agencies became increasingly active in the science of nutrition.

In the United States, the Office of Dietary Supplements of the NIH recommends "average daily recommended amounts" for vitamins and other nutrients. For vitamin C, the basis for the recommendation is a report issued by the National Academy of Medicine in 2000.[6] The report based its recommendations substantially on the Levine studies and the assumption that antioxidants are good for you and the body stores should be kept filled. There are no health outcome data showing that this assumption is correct.[7] For those readers interested in the technical details and the primary literature, the report is available for download from the National Academy Press at www.nap.edu.

The NIH-recommended daily amounts of vitamin C vary by age and physiology: forty milligrams per day for neonates, seventy-five milligrams for adult women, ninety milligrams for adult men. Pregnant and breastfeeding women are advised to ingest up to 120 milligrams per day and smokers an additional 35 milligrams per day because they are subject to increased oxidative stress. Since the median intake of vitamin C from diet alone is on the order of one hundred milligrams per day in the United States (somewhat less in Canada), these guidelines recommend no additional supplementation for most adults. If one were to follow the advice of the US Department of Agriculture and the National Cancer Institute and eat five servings of fresh fruits and vegetables per day, one would consume about two hundred milligrams of vitamin C per day.

The British National Health Service (NHS) is less willing to assume that more antioxidants are necessarily better and summarizes its recommendations more succinctly than the NIH: "Adults (nineteen to sixty-four years) need forty milligrams of vitamin C a day. You should be able to get all the vitamin C you need from your daily diet."[8] This recommendation continues to be based on the amount required to prevent scurvy with a wide margin for error. The two

sides of the Atlantic differ up to threefold in their recommendations. Across the Channel, most European health agencies, like the
United States, recommend about one hundred milligrams daily.

FIGHTING OXYGEN

Epidemiological surveys of populations have correlated higher intakes of fruits and vegetables with a lower incidence of cancer, and
laboratory studies suggest that oxidative stress can increase the ma-
lignant transformation of cells. As a result, antioxidant supplements
became trendy during the 1990s. Vitamins C and E, selenium, and
carotenoids, the orange pigment in carrots, are the favorites.

My friend Bill was part of that trend. He took vitamin C pills
daily when he ran marathons, hoping their antioxidant properties
would help preserve his muscles subjected to the stress of intense
training. He represents one type of consumer who reasons that
supplemental vitamins may do some good and will not do any
harm—a dangerous assumption—and who does not mind wasting
the money if he or she is wrong. Bill took the vitamin pills with little
conviction and eventually stopped.

Mere epidemiologic correlation of fruits and vegetable consumption with cancer incidence is far from definitive proof that
antioxidant intake is beneficial. There are many other constituents
of fruits and vegetables that may contribute to health. If one eats a
lot of fruits and vegetables, one eats less of other foods, such as saturated fats, which may harm one's health. Also, people who consume
diets rich in fruits and vegetables are likely to be of a different so-
cioeconomic group than those who do not. They may be less likely
to have occupational exposure to harmful substances, less likely to
abuse alcohol or drugs, and have better access to medical care.

To gather more definitive evidence, investigators undertook
controlled studies of antioxidants, hoping they would decrease the

incidence of cancer and heart disease. The results have been disappointing. None of the trials individually has shown a clear benefit.

When there are multiple similar studies of a question, researchers try to take maximum advantage of the data by performing what is known as a meta-analysis. The data from all the studies are combined as if it was one large experiment, and the combined data are statistically analyzed. Sometimes effects too small to be detected by the individual studies can be detected by pooling the data, and the method can determine to what extent the studies agree. The results of meta-analyses must be taken with a grain of salt since the individual studies always differ, sometimes in subtle but important ways, in their geographical location, patient populations, and methodology. Frequently the datasets of the individual studies are not available, greatly limiting the strength of any conclusions.

The meta-analysis of antioxidant studies fails to support any benefit of various antioxidant cocktails in the incidence of cancer, heart disease, or cataracts. Even worse news is that there may be a slight increase in mortality in the groups taking the supplemental antioxidants.[9] Therefore, unless there is some combination of antioxidants that shows benefit in future studies, we are dependent on the body's own mechanisms for dealing with oxidizing agents. Disappointingly, it is even possible that taking supplemental antioxidants does more harm than good.

Especially disappointing is the finding that antioxidants do not prevent gastrointestinal cancer.[10] One might hope that the gastrointestinal tract would be an organ where antioxidants might exert a beneficial effect, since they can act on the cells lining the stomach and intestine while passing through the gastrointestinal tract without having to cross the intestinal barrier to reach the systemic circulation. However, randomized trials have found no benefit of supplemental antioxidants. In fact, one analysis showed that a daily intake of one hundred milligrams per day of vitamin C appears to be associated with the lowest incidence of gastric cancer, implying that one cannot do better than eat a reasonably balanced diet.[11]

The other complication is that high concentrations of ascorbic acid can be prooxidant, not antioxidant. The mechanisms regulating the absorption and excretion of oral ascorbic acid ensure that prooxidant concentrations cannot be reached in the circulation except with intravenous administration of high doses. However, there is no such control on the concentration within the gastrointestinal tract, since you can swallow as much as you wish. When you take megadoses, the unabsorbed vitamin C passing through the gut may not fight oxygen; instead, it may help it to damage cells.

FIGHTING DISEASE

Whether vitamin C supplementation either prevents or shortens the common cold continues to linger in the literature. Pharmacies sell a variety of vitamin C concoctions promising stimulation of the immune system and implying, but not specifically claiming, a benefit for colds and flu. Controlled studies have consistently shown that taking supplemental vitamin C does not decrease the incidence of upper respiratory infections in the general, healthy population. However, there is a modest beneficial effect on the symptoms. On average, daily vitamin C supplementation reduces the duration of symptoms by up to 8 percent in adults and 14 percent in children, but only if taken every day, not just when the symptoms of a cold begin.[12]

Studies have also shown a benefit in reducing upper respiratory symptoms in special circumstances. Clinical trials involving ultra-marathon runners in South Africa, military recruits serving in polar environments, children at a ski camp in the Alps, and competitive swimmers have found a modest reduction in the severity of upper respiratory symptoms in groups taking supplemental vitamin C in doses of six hundred to two thousand milligrams per day compared to those taking a placebo.[13] Like the other studies of upper respiratory illness, the vitamin C in these studies was taken every day, not just when symptoms arose, and the benefit was small. Whether it is worth taking a pill twice a day with unknown long-term consequences to obtain this small benefit is an individual decision.

One limitation of studies of the common cold, which is the result of a viral infection, is that the studies depend on the subjects' reports of upper respiratory symptoms rather than direct evidence of a virus. However, one can have a sore throat, cough, or runny nose for reasons other than a viral infection. Interestingly, the situations in which vitamin C appears to benefit upper respiratory symptoms are those in which one might expect the upper respiratory tract to be irritated by inhaling dry air or bathed with chlorinated water.

Perhaps these trials were not studying the common cold but throat irritation. It is unlikely that a pill passing through the throat during a swallow would release enough vitamin to have a local effect, but one might speculate that the temporary spike in plasma levels of vitamin C after a dose might exert some soothing effect on the throat. This might also explain why vitamin C supplementation has a modest effect in reducing the symptoms, but not the incidence, of the common cold. This is pure speculation, and speculation feeds pseudoscience, while science is based on evidence systematically gathered and analyzed.

My friend Mary dismisses the systematic evidence. She is convinced that vitamin C helps her stave off upper respiratory infections. What is her experience? She takes the pills a few times a year when she develops upper respiratory symptoms. Living in New England, she develops such symptoms frequently in the cold, dry winter air. Since she has only one or two colds annually, she reasons that the tablets work more often than not.

North Americans suffer, on average, only one or two colds per year taking no extra vitamins. Mary would get the same number of colds whether she took the pills or not, and she has not tested what would happen if she did not take them. Mary's reasoning is an example of a universal feature of human thought known as confirmation bias. Every time she takes vitamin C and does not get a cold, it reinforces her belief in the pills. Every time she gets a cold despite the extra vitamin C, she passes it off. We are all hard wired to take the same mental shortcut; we pay attention to evidence that supports our prior beliefs and ignore evidence to the contrary.

Confirmation bias leads her to trust her personal experience more than scientific data. She therefore dismisses the multiple controlled experiments showing no benefit of vitamin C taken at the onset of cold symptoms. She also does not realize that all the supplemental vitamin C she takes merely goes down the toilet. Like Pauling, she is a believer. She is no more irrational than any of us. After all, four out of five times she takes the pills, she does not get a cold.

Studies of vitamin C's effects on cancer and heart disease have been similarly disappointing. The studies have showed no clear benefit, further undermining Linus Pauling's promotion of vitamin C as a wonder drug. Readers interested in the details may consult thorough and sympathetic reviews of vitamin C and infections, cancer, and cardiovascular disease.[14]

IS VITAMIN C BAD FOR YOU?

Can you take too much vitamin C? The limits on the absorption of the vitamin and its rapid excretion when body stores are filled are adequate to protect us from systemic toxicity in the short run. However, the vitamin can have effects within the intestine, independent of systemic absorption. The NIH says, "Taking too much vitamin

C can cause diarrhea, nausea, and stomach cramps. In people with a condition called hemochromatosis, which causes the body to store too much iron, high doses of vitamin C could worsen iron overload and damage body tissues." The NIH lists the maximum safe daily amounts varying from four hundred milligrams per day in children one to three years old to two thousand milligrams per day in adults. However, not everyone develops gastrointestinal symptoms, even at a dose of two thousand milligrams per day, and for those who do, the symptoms resolve quickly by reducing the dose.

Another concern is the possibility that megadose vitamin C may increase the risk of kidney stones.[15] In some studies, high doses of vitamin C increase the urinary excretion of uric acid and oxalic acid, two common components of kidney stones. In practice, vitamin C supplementation does not appreciably increase the risk of kidney stones except in people with preexisting kidney disease. A group that may be at particular risk from side effects of vitamin C are those with a genetic deficiency of the enzyme glucose-6-phosphate dehydrogenase, a relatively common condition, especially among those of African or Asian descent. However, toxicity has been reported only after high-dose, intravenous administration, when ascorbic acid can become a prooxidant.[16]

For generally healthy people, even extremely high doses of oral vitamin C appear to be safe, at least for up to three months. The caveat is that without appropriate studies, the long-term effects of chronic administration of megadose vitamin C, both good and bad, are unknown.

YOUR MOTHER WAS RIGHT: EAT YOUR FRUITS AND VEGETABLES

What is the bottom line? By now, it can be concluded that the author is not a proponent of supplemental vitamin C except perhaps a modest amount, fifty milligrams per day, for those with diets deficient in fruits and green vegetables and perhaps for smokers and pregnant women. If one believes antioxidants are beneficial, taking more than two hundred milligrams of ascorbic acid at a time makes little sense. Furthermore, any excess above one hundred milligrams per day, ingested either in the diet or as a pill, will end up in the sewer with the rats who do not need it. They make their own.

As far as is known, supplemental vitamin C is safe, at least in the short term, up to doses of several grams per day. One concern is that studies of supplemental antioxidants have suggested harm with chronic administration, although vitamin C has not been adequately studied in this regard. Two groups should not take vitamin C: people with chronic kidney disease (except those on dialysis) and people with hemochromatosis.

I am a fan of evidence-based medicine. This demands that to justify that a healthy person take any substance—a drug, vitamin, or supplement—well-conducted, controlled studies must show a clear benefit for an important health outcome. An important health outcome means that you feel better, function better, or live longer. If there is no benefit, there are only two other possibilities. The substance will do harm, or it will do nothing and waste money. At times, especially when facing life-threatening disease, one may have no choice but to try unproven therapy. When there is a choice, my advice is to demand unequivocal evidence of benefit. Such evidence does not exist for supplemental vitamin C.

Do what your mother told you: eat a varied diet, including green vegetables and fresh fruit. If, for some reason, that is not possible, take a modest supplement of fifty milligrams of ascorbic acid per day. Taking additional vitamin C certainly will have detrimental effects on your pocketbook with scant evidence that it will benefit your health.

APPENDIX: SELECTED FOOD SOURCES OF VITAMIN C

Food	Milligrams per Serving[1]
Red pepper, sweet, raw, ½ cup	95
Orange juice, ¾ cup	93
Orange, 1 medium	70
Grapefruit juice, ¾ cup	70
Kiwifruit, 1 medium	64
Green pepper, sweet, raw, ½ cup	60
Broccoli, cooked, ½ cup	51
Strawberries, fresh, sliced, ½ cup	49
Brussels sprouts, cooked, ½ cup	48
Grapefruit, ½ medium	39
Broccoli, raw, ½ cup	39
Tomato juice, ¾ cup	33
Cantaloupe, ½ cup	29
Cabbage, cooked, ½ cup	28
Cauliflower, raw, ½ cup	26
Potato, baked, 1 medium	17
Tomato, raw, 1 medium	17
Spinach, cooked, ½ cup	9
Green peas, frozen, cooked, ½ cup	8

NOTES

INTRODUCTION

1. Global News Wire estimates the size of the market for ascorbic acid ("Global Ascorbic Acid Market Poised to Surge from USD 820.4 Million in 2015 to USD 1083.8 Million by 2021," August 24, 2016, https://globe newswire.com/news-release/2016/08/24/866422/0/en/Global-Ascorbic -Acid-Market-Poised-to-Surge-from-USD-820-4-Million-in-2015 -to-USD-1083-8-Million-by-2021-MarketResearchStore-Com.html).

CHAPTER 1. A DISEASE OF MARINERS

1. Quoted in William Byron, *Cervantes: A Biography* (Garden City, NY: Doubleday, 1978), 115.

2. The history of Portuguese exploration and the events leading up to and surrounding the voyage of da Gama are from Roger Crowley, *Conquerers: How Portugal Forged the First Global Empire* (New York: Random House, 2015).

3. All quotations and descriptions of the voyage of Vasco da Gama are from Ernest George Ravenstein, ed., *A Journal of the First Voyage of Vasco de Gama 1497–1499* (1898; repr., London: Hakluyt Society, 2017).

4. Ravenstein, *A Journal of the First Voyage of Vasco da Gama*, 26.

5. Kenneth J. Carpenter, *The History of Scurvy and Vitamin C* (Cambridge: Cambridge University Press, 1986), 1.

6. Stephen R. Bown, *Scurvy: How a Surgeon, a Mariner, and a Gentleman Solved the Greatest Medical Mystery of the Age of Sail* (New York: Thomas Dunne, 2003), 18.

7. Richard Hawkins, *The Observations of Sir Richard Hawkins, Knight, in His Voyage into the South Sea in the Year 1593* (London: Hakluyt Society, 1847; San Francisco: Elibron Classics, 2005).

8. Hawkins, *The Observations of Sir Richard Hawkins*, 82.

9. William Dalrymple, *The Anarchy* (New York: Bloomsbury, 2019), 5.

10. Sir Clements R. Markham, ed., *The Voyages of Sir James Lancaster, KT., to the East Indies* (Miami: Hard Press, 2014).

11. Markham, *The Voyages of Sir James Lancaster*, 62.

12. Carpenter, *The History of Scurvy*, 21.

13. C. Lloyd and J. L. S. Coulter, *Medicine and the Navy, 1200–1900* (Edinburgh: E. and S. Livingstone, 1961).

14. Robert Hitchinson, *The Spanish Armada* (New York: Thomas Dunne, 2014), 203–4.

15. Carpenter, *The History of Scurvy*, 7–10.

CHAPTER 2. CATASTROPHE AND ENLIGHTENMENT

1. Stephen R. Bown, *Scurvy: How a Surgeon, a Mariner, and a Gentleman Solved the Greatest Medical Mystery of the Age of Sail* (New York: Thomas Dunne, 2003), 68.

2. The account of Anson's voyage is from the journal published under his name: George Anson, *A Voyage around the World*, ed. Richard R. Walters (London: Paean, 2011). Also see Bown, *Scurvy*, 47–69.

3. Arthur Herman, *How the Scots Invented the Modern World* (New York: Crown, 2001).

4. Louis H. Roddis, *James Lind: Founder of Nautical Medicine* (New York: Henry Schuman, 1950); Ralph Stockman, "James Lind and Scurvy," *Edinburgh Medical Journal* 33 (1926): 329–50.

5. James Lind, *A Treatise on the Scurvy*, 3rd ed. (London, 1772; repr., Birmingham, AL: Classics in Medicine Library, 1980), 150.

6. Quoted by Bown, *Scurvy*, 82.

7. Lind, *A Treatise on the Scurvy*, 243.

8. Lind, *A Treatise on the Scurvy*, 522.

CHAPTER 3. AN UNLIKELY HERO
AND A PARTIAL VICTORY

1. Francis E. Cuppage, *James Cook and the Conquest of Scurvy* (Westport, CT: Greenwood Press, 1994).

2. Stephen R. Bown, *Scurvy: How a Surgeon, a Mariner, and a Gentleman Solved the Greatest Medical Mystery of the Age of Sail* (New York: Thomas Dunne, 2003), 166.

3. J. C. Beaglehole, *The Life of Captain James Cook* (Stanford, CA: Stanford University Press, 1974).

4. J. Cook, "The Methods Taken for Preserving the Health of the Crew of His Majesty's Ship the *Resolution* during Her Late Voyage around the World," *Philosophical Transactions of the Royal Society London* 66 (1776): 402–6; B. J. Stubbs, "Captain Cook's Beer: The Antiscorbutic Use of Malt and Beer in Late 18th Century Sea Voyages," *Asia Pacific Journal of Clinical Nutrition* 13 (2003): 129–37.

5. R. D. Leach, "Sir Gilbert Blane, Bart, MD FRS (1749–1832)," *Annals of the Royal College of Surgeons of England* 62 (1980): 232–39; J. G. Penn-Barwell, "Sir Gilbert Blane FRS: The Man and His Legacy," *Journal of the Royal Naval Medical Service* 102 (2016): 61–66; M. Wharton, "Sir Gilbert Blane Bt (1749–1834)," *Annals of the Royal College of Surgeons of England* 66 (1984): 375–76.

6. Gilbert Blane, *Observations on the Diseases of Seamen*, 2nd ed. (London: Joseph Cooper, 1789), 92, Kindle.

7. Blane, *Observations of the Diseases of Seamen*, 2,567.

8. Blane, *Observations on the Diseases of Seamen*.

9. Blane, *Observations of the Diseases of Seamen*, 611.

10. Blane, *Observations of the Diseases of Seamen*, 625.

11. Blane, *Observations of the Diseases of Seamen*, 1,750.

12. Blane, *Observations of the Diseases of Seamen*, 1,750.

13. Steven Johnson, *The Ghost Map* (New York: Riverhead, 2007).

14. Leach, "Sir Gilbert Blane."

15. James Dugan, *The Great Mutiny* (New York: G. P. Putnam's Sons, 1965), 56.

16. Dugan, *The Great Mutiny*.

17. Dugan, *The Great Mutiny*, 103.

CHAPTER 4. STEPS FORWARD AND BACK

1. Christopher Lloyd, "The Introduction of Lemon Juice as a Cure for Scurvy," *Bulletin of the History of Medicine* 35 (1961): 123–32; J. H. Baron, "Sailors' Scurvy before and after James Lind—A Reassessment," *Nutrition Reviews* 67 (2009) 315–32; M. Harrison, "Scurvy on Sea and Land: Political Economy and Natural History, c. 1780–c. 1850," *Journal for Maritime Research* 15 (2013): 7–25.

2. H. Chick, "The Discovery of Vitamins," *Progress in Food and Nutrition Science* 1 (1975): 1–20.

3. G. C. Cook, "Scurvy in the British Mercantile Marine in the 19th Century, and the Contribution of the Seamen's Hospital Society," *Postgraduate Medical Journal* 80 (2017): 224–29.

4. Cook, "Scurvy in the British Mercantile Marine."

5. David I. Harvie, *Limeys: The True Story of One Man's War against Ignorance, the Establishment and the Deadly Scurvy* (Stroud, UK: Sutton Publishing, 2002), 215–24.

6. Kenneth J. Carpenter, *The History of Scurvy and Vitamin C* (Cambridge: Cambridge University Press, 1986), 98–132.

7. Chick, "The Discovery of Vitamins."

8. R. Christison, "On Scurvy. Account of Scurvy as It Has Lately Appeared in Edinburgh, and of an Epidemic of It among Railway Labourers in the Surrounding County," *Monthly Journal of Medical Science* 13, no. 74 (1847): 1–22.

9. M. Harrison, "Scurvy on Sea and Land: Political Economy and Natural History, c. 1780–c. 1850," *Journal for Maritime Research* 15 (2013): 7–25.

10. W. Baly, "On the Prevention of Scurvy in Prisoners, Pauper Lunatic Asylums, Etc.," *London Medical Gazette* 1 (1843): 699–703.

11. Cecil Woodham-Smith, *The Great Hunger: Ireland 1845–1849* (New York: E. P. Dutton, 1980).

12. P. P. Boyle and C. O. Grada, "Fertility Trends, Excess Mortality and the Great Irish Famine," *Demography* 23, no. 4 (1986): 543–62.

13. C. Ritchie, "Contributions to the Pathology and Treatment of the Scorbutus, Which Is at Present Prevalent in Various Parts of Scotland," *Monthly Journal of Medical Science* 12, no. 13 (1847): 38–49.

14. Carpenter, *The History of Scurvy*, 123–26.

15. R. K. Aspin, "The Papers of Sir Thomas Barlow, BT, KVCO, FRS, PRCP (1845–1945)," *Medical History* 37 (1993): 333–40.

16. T. Barlow, "On Cases Described as 'Acute Rickets' Which Are Probably a Combination of Scurvy and Rickets, the Scurvy Being an Essential, and the Rickets a Variable, Element," *Medical and Chirurgical Transactions* 66 (1883): 159–220.

17. T. Barlow, "The Bradshaw Lecture on Infantile Scurvy and Its Relation to Rickets," *British Medical Journal* 2, no. 1767 (1894): 1029–34.

18. A. F. Hess and M. Fish, "Infantile Scurvy: The Blood, the Blood-Vessels and the Diet," *American Journal of Diseases of Children* 8 (1914): 385–105.

19. Carpenter, *The History of Scurvy*, 134.

20. Stephen R. Bown, *Scurvy: How a Surgeon, a Mariner, and a Gentleman Solved the Greatest Medical Mystery of the Age of Sail* (New York: Thomas Dunne, 2003), 82–83.

21. A. H. Smith, "A Historical Inquiry into the Efficacy of Lime-Juice for the Prevention and Cure of Scurvy," *Journal of the Royal Army Medical Corps* 32 (1919): 93–116, 188–208; L. G. Wilson, "The Clinical Definition of Scurvy and the Discovery of Vitamin C," *Journal of the History of Medicine and Allied Science* 30 (1975): 40–60.

22. Carpenter, *The History of Scurvy*, 145.

23. E. A. Wilson, "The Medical Aspect of the *Discovery*'s Voyage to the Antarctic," *British Medical Journal* 2, no. 2323 (1905): 77–80.

24. Edward A. Wilson, *Diary of the "Discovery" Expedition* (London: Blanford Press, 1966), 287.

25. Wilson, "The Medical Aspect of the *Discovery*'s Voyage to the Antarctic."

26. Roland Huntford, *The Last Place on Earth* (New York: Modern Library, 1999).

27. Wilson, "The Clinical Definition of Scurvy and the Discovery of Vitamin C."

CHAPTER 5. A DIFFERENT KIND OF NUTRIENT

1. Elmer V. McCollum, *A History of Nutrition* (Boston: Houghton Mifflin, 1957), 75–81.

2. F. G. Hopkins, "The Earlier History of Vitamin Research," Nobel lecture, December 11, 1929, www.nobelprize.org/prizes/medicine/1929 /hopkins/lecture/; R. D. Semba, "The Discovery of the Vitamins," *International Journal for Vitamin and Nutrition Research* 82 (2012): 310–15.

3. Kenneth J. Carpenter, *Beriberi, White Rice and Vitamin B* (Berkeley: University of California Press, 2000), 10–14; K. C. Carter, "The Germ Theory, Beriberi, and the Deficiency Theory of Disease," *Medical History and Bioethics* 21 (1977): 119–36; Semba, "The Discovery of the Vitamins."

4. Medical Research Committee Special Report No. 20, *Report on the Present State of Knowledge Concerning Accessory Food Factors (Vitamines)* (London: His Majesty's Stationery Office, 1919).

5. Carpenter, *Beriberi*, 10–14.

6. A. Bay, "Mori Ōgai Mori and the Beriberi Dispute," *East Asian Science, Technology and Society: An International Journal* 5 (2011): 573–779.

7. Carpenter, *Beriberi*, 35–46.

8. C. Eijkman, "Antineuitic Vitamin and Beriberi," Nobel lecture, 1929, www.nobelprize.org/prizes/medicine/1929/eijkman/lecture/.

9. G. Grijns, "Over Polyneuritis Gallinarum," *Geneeskundig Tijdschrift voor Nererlandsch-Indië* 41 (1901): 3–110. Published in English in G. Grijns, *Researches on Vitamins 1900–1911* (Gorinchem: J. Noorduyn en Zoon N.V., 1935), 1–108.

10. Eijkman "Antineuitic Vitamin and Beriberi."

11. Thomas Kuhn, *The Structure of Scientific Revolutions*, 2nd ed. (Chicago: University of Chicago Press, 1970).

12. A. Holst, "Experimental Studies Relating to 'Ship Beri-Beri' and Scurvy," *J Hygiene* 7 (1907): 619–33. A. Holst, and T. Frølich "Experimental Studies Relating to Ship Beri-Beri and Scurvy: II. On the Etiology of Scurvy." *Journal of Hygiene* 7 (1907): 634–71.

13. Elmer V. McCollum, *A History of Nutrition* (Boston: Houghton Mifflin, 1957), 201–28.

14. E. V. McCollum and W. Pitz, "The 'Vitamine' Hypothesis and Deficiency Diseases," *Journal of Biological Chemistry* 31 (1917): 229–53.

15. F. G. Hopkins, "Feeding Experiments Illustrating the Importance of Accessory Factors in Normal Dietaries," *Journal of Physiological Sciences* 44 (1912): 425–60.

16. A. Maltz, "Casimer Funk, Nonconformist Nomenclature, and Networks Surrounding the Discovery of Vitamins," *Journal of Nutrition* 143 (2013): 1013–20; A. Piro, G. Tagarelli, P. Lagonia, A. Tagarelli, and A. Quattrone, "Casimer Funk: His Discovery of the Vitamins and Their Deficiency Disorders," *Annals of Nutrition and Metabolism* 57 (2010): 85–88.

17. C. Funk, "On the Chemical Nature of the Substance Which Cures Polyneuritis in Birds Induced by a Diet of Polished Rice," *Journal of Physiology* 43 (1911): 395–400.

18. J. C. Drummond, "Note on the Role of the Anti-Scorbutic Factor in Nutrition," *Biochemical Journal* 13 (1919): 77–80.

19. Patricia Fara, *A Lab of One's Own: Science and Suffrage in the First World War* (Oxford: Oxford University Press, 2018).

20. Lynn Brindan, Alison Brading, and Tilli Taney, eds., *Women Physiologists: An Anniversary Celebration of their Contributions to British Physiology* (London: Portland Press, 1993).

21. H. Chick, E. M. Hume, R. F. Skelton, and A. Henderson Smith, "The Relative Content of Antiscorbutic Principle in Limes and Lemons," *Lancet* (November 30, 1918): 735–38.

22. A. Henderson Smith, "A Historical Inquiry into the Efficacy of Lime-Juice for the Prevention and Cure of Scurvy," *Journal of the Royal Army Medical Corps* (1919): 188–208.

23. H. Chick, E. J. Dalyell, M. Hume, H. M. M. Mackay, and A. Henderson Smith, "The Etiology of Rickets in Infants," *Lancet* 2 (1922): 7–11.

24. E. V. McCollum and M. Davis, "The Necessity of Certain Lipins in the Diet during Growth," *Journal of Biological Chemistry* 15 (1913): 167–75.

25. McCollum and Pitz, "The 'Vitamine' Hypothesis and Deficiency Diseases."

26. McCollum and Davis, "The Necessity of Certain Lipins in the Diet during Growth."

27. Drummond, "Note on the Role of the Anti-Scorbutic Factor in Nutrition."

28. Drummond, "Note on the Role of the Anti-Scorbutic Factor in Nutrition."

29. F. G. Hopkins, "The Analyst and the Medical Man," *Analyst* 31 (1906): 385–404.

CHAPTER 6. THE VITAMIN HUNTERS

1. S. S. Zilva, "The Isolation and Identification of Vitamin C," *Archives of Disease in Childhood* 10 (1935): 253–64.

2. A. Harden and S. S. Zilva, "The Antiscorbutic Factor in Lemon Juice," *Biochemical Journal* 12 (1918): 259–69.

3. Ralph W. Moss, *Free Radical: Albert Szent-Gyorgyi and the Battle over Vitamin C* (New York: Paragon House, 1988).

4. A. Szent-Gyorgyi, "Lost in the Twentieth Century," *Annual Review of Biochemistry* 32 (1963): 1–15.

5. A. Szent-Gyorgyi, "Observations on the Function of Peroxidase Systems and the Chemistry of the Adrenal Cortex," *Biochemical Journal* 22 (1928): 1387–1410.

6. J. L. Svirbely and C. G. King, "The Preparation of Vitamin C Concentrates from Lemon Juice," *Journal of Biological Chemistry* 94 (1931): 483–90.

7. J. L. Svirbely and A. Szent-Gyorgyi, "Hexuronic Acid as the Antiscorbutic Factor," *Nature* 129 (1932): 576; J. L. Svirbely and A. Szent-Gyorgyi, "The Chemical Nature of Vitamin C," *Biochemical Journal* 26 (1932): 865–70.

8. S. S. Zilva, "Hexuronic Acid as the Antiscorbutic Factor," *Nature* 129 (1932): 943; S. S. Zilva, "The Isolation and Identification of Vitamin C."

9. Szent-Gyorgyi, "Lost in the Twentieth Century."

10. C. G. King and W. A. Waugh, "The Chemical Nature of Vitamin C," *Science* 75 (1932) 357–58.

11. W. A. Waugh and C. G. King, "Isolation and Characterization of Vitamin C," *Journal of Biological Chemistry* 97 (1932): 325–31.

12. Svirbely and Szent-Gyorgyi, "Hexuronic Acid as the Antiscorbutic Factor."

13. Szent-Gyorgyi, "Lost in the Twentieth Century."

14. W. N. Haworth, "The Structure of Carbohydrates and of Vitamin C," in *Nobel Lectures: Chemistry 1922–1941* (Amsterdam: Elsevier, 1966).

15. A. Szent-Gyorgyi and W. N. Haworth, "Hexuronic Acid (Ascorbic Acid) as the Antiscorbutic Factor," *Nature* 131 (1933): 24.

16. A. Szent-Gyorgyi, "Oxidation, Energy Transfer, and Vitamins," in *Nobel Lectures: Physiology or Medicine 1922–1941* (Amsterdam: Elsevier, 1965); W. N. Haworth, "The Structure of Carbohydrates and of Vitamin C," in *Nobel Lectures: Chemistry 1922–1941* (Amsterdam: Elsevier, 1966).

17. G. J. Cox, "Crystallized Vitamin C and Hexuronic Acid," *Science* 86 (1937): 540–42.

CHAPTER 7. SCURVY FOR SCIENCE

1. J. C. Drummond and A. Wilbraham, "William Stark, M.D.," *Lancet* 226 (1935): 459–62.

2. Adrian Tinniswood, *The Royal Society and the Invention of Modern Science* (New York: Basic, 2019), 78.

3. J. H. Crandon, C. C. Lund, and D. B. Dill, "Experimental Human Scurvy," *New England Journal of Medicine* 223 (1940): 353–69; J. H. Crandon and C. C. Lund, "Vitamin C Deficiency in an Otherwise Normal Adult," *New England Journal of Medicine* 222 (1940): 748–52.

4. Lawrence K. Altman, *Who Goes First? The Story of Self-Experimentation in Medicine* (New York: Random House, 1987), 250–55.

5. J. Pemberton, "Medical Experiments Carried out in Sheffield on Conscientious Objectors to Military Service during the 1939–45 War," *International Journal of Epidemiology* 35 (2006): 556–58; Medical Research Council, *Vitamins: A Survey of Present Knowledge*, Medical Research Council Special Reports Series No. 167 (London: His Majesty's Stationery Office, 1932).

6. M. Pijoan and E. L. Lozner, "Vitamin C Economy in the Human Subject," *Bulletin of the Johns Hopkins Hospital* 75 (1944): 303–14.

7. R. E. Hodges, E. M. Baker, J. Hood, H. E. Sauberlich, and S. E. March, "Experimental Scurvy in Man," *American Journal of Clinical Nutrition* 22 (1969): 535–48; R. E. Hodges, J. Hood, J. E. Canham, H. E. Sauberlich, and E. M. Baker, "Clinical Manifestations of Ascorbic Acid Deficiency in Man," *Amerian Journal of Clinical Nutrition* 24 (1971): 432–43.

8. S. K. Shah, F. G. Miller, D. C. Darton, D. Duenas, C. Emerson, H. Fernandez Lynch, E. Jamrozik, N. S. Jecker, D. Kamuya, M. Kapulu, J. Kimmelman, D. Mackay, M. J. Memoli, S. C. Murphy, R. Palacios, T. L. Richie, M. Roestenberg, A. Saxena, K. Saylor, M. J. Selgelid, V. Vaswani, and A. Rid, "Ethics of Controlled Human Infection to Address COVID-19," *Science* 368 (2020): 832–34.

CHAPTER 8. NORMAL SCIENCE

1. C. S. Johnston, F. M. Steinberg, and R. B. Rucker, "Ascorbic Acid," in *Handbook of Vitamins,* ed. J. Zempleni, 4th ed. (Boca Raton, FL: CRC Press, 2007), 489–520.

2. M. Levine, "New Concepts in the Biology and Biochemistry of Ascorbic Acid," *New England Journal of Medicine* 314 (1986): 892–902; I. B. Chatterjee, A. K. Mujumder, B. K. Nandi, and N. Subramanian,

"Synthesis and Some Major Functions of Vitamin C in Animals," *Annals of the New York Academy of Sciences* 258 (1975): 24–47; J. Mandl, A. Szarka, and G. Banhegyi, "Vitamin C: Update on Physiology and Pharmacology," *British Journal of Pharmacology* 157 (2009) 1097–1110; Johnston, Steinberg, and Rucker, "Ascorbic Acid."

3. Mandl, Szarka, and Banhegyi, "Vitamin C: Update on Physiology and Pharmacology."

4. Johnston, Steinberg, and Rucker, "Ascorbic Acid."

5. N. L. Parrow, J. A. Leshin, and M. Levine, "Parenteral Ascorbate as a Cancer Therapeutic: A Reassessment Based on Pharmacokinetics," *Antioxidants and Redox Signaling* 19 (2013): 2141–56.

6. N. Smirnoff, "Ascorbic Acid Metabolism and Function: A Comparison of Plants and Animals," *Free Radical Biology and Medicine* 122 (2018): 116–29; G. Drouin, J. R. Godin, and B. Page, "The Genetics of Vitamin C Loss in Vertebrates," *Current Genomics* 12 (2011): 371–78; I. B. Chatterjee, "Evolution and the Biosynthesis of Ascorbic Acid," *Science* 182 (1973): 1271–72; A. Nandi, K. Mukhopadhyay, M. K. Ghosh, D. J. Chattopadhyay, and I. B. Chatterjee, "Evolutionary Significance of Vitamin C Biosynthesis in Terrestrial Vertebrates," *Free Radical Biology and Medicine* 22 (1997): 1047–54; A. R. Fernie and T. Tohge, "Ascorbate Biosynthesis: A Cross-Kingdom History," *eLife* 4 (2015): e07527.

7. P. Aghajanian, S. Hall, M. D. Wongworawat, and S. Mohan, "The Roles and Mechanisms of Action of Vitamin C in Bone: New Developments," *Journal of Bone and Mineral Research* 30 (2015): 1945–55; Johnston, Steinberg, and Rucker, "Ascorbic Acid."

8. N. Gest, H. Gaitier, and R. Stevens, "Ascorbate as Seen through Plant Evolution: The Rise of a Successful Molecule?" *Journal of Experimental Botany* 64 (2013): 33–53; B. N. Ivanov, "Role of Ascorbic Acid in Photosynthesis," *Biochemistry (Moscow)* 79 (2014): 282 89; Y. Leshem, "Plant Senescence Processes and Free Radicals," *Free Radical Biology and Medicine* 5 (1988): 39–49; Smirnoff, "Ascorbic Acid Metabolism and Function: A Comparison of Plants and Animals"; N. Smirnoff and G. L. Wheeler, "Ascorbic Acid in Plants: Biosynthesis and Function," *Critical Reviews in Biochemistry and Molecular Biology* 35 (2000): 291–414.

9. Mandl, Szarka, and Banhegyi, "Vitamin C: Update on Physiology and Pharmacology."

NOTES

10. S. England and S. Seifter, "The Biochemical Functions of Ascorbic Acid," *Annual Review of Nutrition* 6 (1986): 365–406.

11. Johnston, Steinberg, and Rucker, "Ascorbic acid."

12. Mandl, Szarka, and Banhegyi, "Vitamin C: Update on Physiology and Pharmacology."

13. K. J. Nytko, N. Maeda, P. Schafli, P. Spielman, R. H. Wengler, and D. P. Stiehl, "Vitamin C Is Dispensable for Oxygen Sensing in Vivo," *Blood* 117 (2010): 5485–93.

14. Johnston, Steinberg, and Rucker, "Ascorbic Acid."

15. S. Hasselhot, P. Tveden-Nyborg, and J. Lykkesfeldt, "Distribution of Vitamin C Is Tissue Specific with Early Saturation of the Brain and Adrenal Glands Following Differential Oral Dose Regimens in Guinea Pigs," *British Journal of Nutrition* 113 (2015): 1539–49.

16. C. C. Carr and S. Maggini, "Vitamin C and Immune Function," *Nutrients* 9 (2017): 1211, https://doi.org/10.3390/nu9111211; H. Hemila, "Vitamin C and Infections," *Nutrients* 9 (2017): 339–56; W. R. Thomas and P. G. Holt, "Vitamin C and Immunity: An Assessment of the Evidence," *Clinical and Experimental Immunology* 32 (1978): 370–79.

17. Alfred F. Hess, *Scurvy: Past and Present* (Philadelphia: Lippencott, 1920). Available at http://chla.library.cornell.edu.

18. P. W. Washko, Y. Wang, and M. Levine, "Ascorbic Acid Recycling in Human Neutrophils," *Journal of Biological Chemistry* 268 (1993): 15531–35.

CHAPTER 9. THE PASSION OF LINUS PAULING

1. The details of the biography of Linus Pauling are from Thomas Hager, *Force of Nature: The Life of Linus Pauling* (New York: Simon and Schuster, 1995).

2. L. Pauling, "The Nature of the Chemical Bond: Application of Results Obtained from the Quantum Mechanics and from a Theory of Paramagnetic Susceptibility to the Structure of Molecules," *Journal of the American Chemical Society* 53 (1931): 1367–400.

3. L. Pauling, R. B. Corey, and H. R. Branson, "The Structure of Proteins: Two Hydrogen-Bonded Helical Configurations of the

Polypeptide Chain," *Proceedings of the National Academy of Sciences* 37 (1951): 205–11; L. Pauling and R. B. Corey, "Atomic Coordinates and Structure Factors for Two Helical Configurations of Polypeptide Chains," *Proceedings of the National Academy of Sciences* 37 (1951): 235–40; L. Pauling and R. B. Corey, "The Structure of Synthetic Polypeptides," *Proceedings of the National Academy of Sciences* 37 (1951): 241–50; L. Pauling and R. B. Corey, "The Pleated Sheet, a New Layer Configuration of Polypeptide Chains," *Proceedings of the National Academy of Sciences* 37 (1951): 251–56; L. Pauling and R. B. Corey, "The Structure of Feather Rachis Keratin," *Proceedings of the National Academy of Sciences* 37 (1951): 256–61; L. Pauling and R. B. Corey, "The Structure of Hair, Muscle and Related Proteins," *Proceedings of the National Academy of Sciences* 37 (1951): 261–71; L. Pauling and R. B. Corey, "The Structure of Fibrous Proteins of the Collagen-Gelatin Group," *Proceedings of the National Academy of Sciences* 37 (1951): 272–81; L. Pauling and R. B. Corey, "The Polypeptide-Chain Configuration in Hemoglobin and Other Globular Proteins," *Proceedings of the National Academy of Sciences* 37 (1951): 282–85; L. Pauling and R. B. Corey, "Configurations of Polypeptide Chains with Favored Orientations around Single Bonds: Two New Pleated Sheets," *Proceedings of the National Academy of Sciences* 37 (1951): 729–40.

4. L. Pauling, H. A. Itano, S. J. Singer, and I. C. Wells, "Sickle Cell Anemia: A Molecular Disease," *Science* 110 (1949): 543–48.

5. L. Pauling and R. B. Corey, "A Proposed Structure for the Nucleic Acids," *Proceedings of the National Academy of Sciences* 39 (1953): 84–97.

6. L. Pauling, "Orthomolecular Psychiatry," *Science* 160 (1968): 265–71.

7. Linus Pauling, *Vitamin C and the Common Cold* (San Francisco: W. H. Freeman, 1970).

8. G. Ritzel, "Critical Evaluation of the Prophylactic and Therapeutic Properties of Vitamin C with Respect to the Common Cold," *Helvetica Medica Acta* 28 (1961): 63–68.

9. L. Pauling, "The Significance of the Evidence about Ascorbic Acid and the Common Cold," *Proceedings of the National Academy of Sciences* 68 (1971): 2678–81.

10. D. W. Cowan, H. S. Diehl, and A. B. Baker, "Vitamins for the Prevention of Colds," *JAMA* 120 (1942): 1268–71; T. R. Karlowski, T. C. Chalmers, L. D. Frenkel, A. Z. Kapikian, T. L. Lewis, and J. M. Lynch, "Ascorbic Acid for the Common Cold: A Prophylactic and Therapeutic Trial," *JAMA* 231 (1975): 1038–42; T. W. Anderson, G. H. Beaton, P. N. Corer, and L. Spero, "Winter Illness and Vitamin C: The Effect of Relatively Low Doses," *Canadian Medical Association Journal* 112 (1975): 823–26; T. W. Anderson, D. B. W. Reid, and G. H. Beaton, "Vitamin C and the Common Cold: A Double-Blind Trial," *Canadian Medical Association Journal* 105 (1972): 503–8; T. W. Anderson, G. Suranyi, and G. H. Beaton, "The Effect on Winter Illness of Large Doses of Vitamin C," *Canadian Medical Association Journal* 111 (1974): 31–36; J. L. Coulehan, K. S. Reisinger, K. D. Rogers, and D. W. Bradley, "Vitamin C Prophylaxis in a Boarding School," *New England Journal of Medicine* 290 (1974): 6–10; C. W. M. Wilson and H. S. Loh, "Common Cold and Vitamin C," *Lancet* 1 (1973): 638–41.

11. T. C. Chalmers, "Effects of Ascorbic Acid on the Common Cold," *American Journal of Medicine* 58 (1975): 532–36.

12. Chalmers, "Effects of Ascorbic Acid on the Common Cold"; M. H. M. Dykes and P. Meier, "Ascorbic Acid and the Common Cold: Evaluation of Its Efficacy and Toxicity," *JAMA* 231 (1975): 1073–79; T. W. Anderson, "Large-Scale Trials of Vitamin C," *Annals of the New York Academy of Sciences* 258 (1975): 498–504.

13. Linus Pauling, *Vitamin C, the Common Cold, and the Flu* (San Francisco: W. H. Freeman, 1976).

14. E. T. Creagan, C. G. Moertel, J. R. O'Fallon, A. J. Schutt, M. J. O'Connell, J. Rubin, and S. Frytak, "Failure of High-Dose Vitamin C (Ascorbic Acid) to Benefit Patients with Advanced Cancer: A Controlled Trial," *New England Journal of Medicine* 301 (1979): 687–90.

15. C. G. Moertel, T. R. Fleming, E. T. Creagan, J. Rubin, M. J. O'Connell, and M. M. Ames, "High-Dose Vitamin C versus Placebo in the Treatment of Patients with Advanced Cancer Who Have Had No Prior Chemotherapy: A Randomized Double-Blind Comparison," *New England Journal of Medicine* 312 (1985): 137–41.

CHAPTER 10. VITAMINS, BUSINESS, AND POLITICS

1. Dan Hurley, *Natural Causes: Death, Lies, and Politics in America's Vitamin and Herbal Supplement Industry* (New York: Broadway, 2006); Paul A. Offit, *Do You Believe in Magic? The Sense and Nonsense of Alternative Medicine* (New York: HarperCollins, 2013).

2. E. D. Kantor, C. D. Rehm, M. Du, E. White, and E. L. Giovannucci, "Trends in Dietary Supplement Use among US Adults from 1999–2012," *JAMA* 316 (2016): 1464–74; D. M. Eisenberg, R. C. Kessler, C. Foster, F. E. Norlock, D. R. Calkins, and T. L. Delbanco, "Unconventional Medicine in the United States," *New England Journal of Medicine* 328 (1993): 246–52; J. J. Galche, R. I. Bailey, N. Potischman, and J. T. Dwyer, "Dietary Supplement Use Was Very High among Older Adults in the United States in 2011–2014," *Journal of Nutrition* 147 (2017): 1968–76; S. P. Murphy, D. Rose, M. Hudes, and F. E. Viterii, "Demographic and Economic Factors Associated with Dietary Quality for Adults in the 1987–88 Nationwide Food Consumption Survey," *Journal of the American Dietetic Association* 92 (1992): 1352–57.

3. D. M. Qato, J. Wilder, P. Shumm, V. Gillet, and C. Alexander, "Changes in Prescription and Over-the-Counter Medication and Dietary Supplement Use among Older Adults in the United States, 2005 versus 2011," *JAMA Internal Medicine* 176 (2016): 473–82.

4. U. S. Food and Drug Administration, "Dietary Supplements," https://www.fda.gov/Food/DietarySupplements/default.htm.

5. S. M. Schmitz, H. L. Lopez, D. Mackay, H. Nguyen, and P. Miller, "Serious Adverse Events Reported with Dietary Supplement Use in the United States: A 2.5 Year Experience," *Journal of Dietary Supplements* 17 (2020): 227–48, https://doi.org/10.1080/19390211.2018.1513109.

6. A. I. Geller, N. Shehab, N. J. Weidle, M. C. Lovegrove, B. J. Wolpert, B. B. Timbo, R. P. Mozersky, and D. S. Budnitz, "Emergency Department Visits for Adverse Events Related to Dietary Supplements," *New England Journal of Medicine* 373 (2015): 1531–40.

7. J. Calahan, D. Howard, A. J. Almalki, M. P. Gupta, and A. I. Calderon, "Chemical Adulteration in Herbal Medicinal Products: A

Review," *Planta Medica* 82 (2016): 505–15; D. M. Marcus, "Dietary Supplements: What's in a Name? What's in the Bottle?" *Drug Testing Analysis* 8 (2015): 410–12.

8. Katherine Eban, *Bottle of Lies* (New York: HarperCollins, 2019).

9. Linus Pauling, *The Nature of the Chemical Bond* (Ithaca, NY: Cornell University Press, 1960).

CHAPTER 11. LESSONS LEARNED

1. Medical Research Council, *Vitamins: A Survey of Present Knowledge*, Medical Research Council Special Reports Series No. 167 (London: His Majesty's Stationery Office, 1932), 10.

2. A. E. Carroll, "Health Facts Aren't Enough. Should Persuasion Become a Priority?" *New York Times*, July 22, 2019.

3. B. Nyhan, J. Reifler, S. Richey, and G. L. Freed, "Effective Messages in Vaccine Promotion: A Randomized Trial," *Pediatrics* 133 (2014): 835–42.

4. R. J. Blendon, C. M. DesRoches, J. M. Benson, M. Brodie, and D. E. Altman, "Americans' Views on the Use and Regulation of Dietary Supplements," *Archives of Internal Medicine* 161 (2001): 805–10.

5. A. J. Bollet, "Politics and Pellagra: The Epidemic of Pellagra in the U.S. in the Early Twentieth Century," *Yale Journal of Biology and Medicine* 65 (1992): 211–21.

CHAPTER 12. A GUIDE FOR THE PERPLEXED

1. Food and Nutrition Board, Institute of Medicine, *Dietary Reference Intakes for Vitamin C, Vitamin E, Selenium and Carentenoids* (Washington, DC: National Academy Press, 2000).

2. Y. Li and H. E. Schelhorn, "New Developments and Novel Therapeutic Perspectives for Vitamin C," *Journal of Nutrition* 137 (2007): 2171–84.

3. M. Levine, C. Conry-Cantelena, Y. Wang, R. W. Welch, P. W. Washko, K. R. Dhariwal, J. B. Park, A. Lazarev, J. F. Graumlich, J. King,

and L. R. Cantilena, "Vitamin C Pharmacokinetics in Healthy Volunteers: Evidence for a Recommended Daily Allowance," *Proceedings of the National Academies of Science* 93 (1996): 3704–9; M. Levine, Y. Wang, S. J. Padayatty, and J. Morrow, "A New Recommended Dietary Allowance of Vitamin C for Healthy Women," *Proceedings of the National Academies of Science* 98 (2001): 9842–46.

4. Food and Nutrition Board, *Dietary Reference Intakes for Vitamin C, Vitamin E, Selenium and Carentenoids.*

5. A. Kallner, D. Hartmann, and D. Hornig, "Steady-State Turnover and Body Pool of Ascorbic Acid in Man," *American Journal of Clinical Nutrition* 32 (1979): 530–39.

6. Food and Nutrition Board, *Dietary Reference Intakes for Vitamin C, Vitamin E, Selenium and Carentenoids.*

7. V. R. Young, "Evidence for a Recommended Daily Allowance for Vitamin C from Pharmacokinetics: A Comment and Analysis," *Proceedings of the National Academies of Science* 93 (1996): 14344–48.

8. Public Health England, *Government Dietary Recommendations* (London: The Stationery Office, 2016), https://assets.publishing.service.gov.uk/government/uploads/system/uploads/attachment_data/file/618167/government_dietary_recommendations.pdf.

9. G. Bjelakovic, D. Nikolova, L. L. Gluud, R. G. Simonetti, and C. Gluud, "Antioxidant Supplements for Prevention of Mortality in Healthy Participants and Patients with Various Diseases," *Cochrane Database of Systematic Reviews* CD007176 (2012).

10. G. Bjelakovic, D. Nikolova, R. G. Simonetti, and C. Gluud, "Antioxidant Supplements for Prevention of Gastrointestinal Cancers: A Systematic Review and Meta-Analysis," *Lancet* 364 (2004): 1219–28.

11. Bjelakovic, Nikolova, Simonetti, and Gluud, "Antioxidant Supplements for Prevention of Gastrointestinal Cancers."

12. H. Hemila and E. Chalker, "Vitamin C for Preventing and Treating the Common Cold," *Cochrane Database of Systematic Reviews* CD000980 (2013).

13. G. Ritzel, "Critical Evaluation of the Prophylactic and Therapeutic Properties of Vitamin C with Respect to the Common Cold," *Helvetica Medica Acta* 28 (1961): 63–68; N. W. Constantini, G. Dubnov-Raz, B. Eyal, E. M. Berry, A. H. Cohen, and H. Hemila, "The Effects of

Vitamin C on Upper Respiratory Infections in Adolescent Swimmers: A Randomized Trial," *European Journal of Pediatrics* 170 (2011): 59–63; B. H. Sabiston and M. W. Radonski, "Health Problems and Vitamin C in Canadian Northern Military Operations," Defence and Civil Institute of Environmental Medicine Report 74-R-1012 (1974): www.mv.helsinki.fi/home/hemila/CC/Sabiston_1974_ch.pdf; E. M. Peters, J. M. Goetzsche, B. Grobbelaar, and T. D. Noakes, "Vitamin C Supplementation Reduces the Incidence of Postrace Symptoms of Upper-Respiratory-Tract Infection in Ultramarathon Runners," *American Journal of Clinical Nutrition* 57 (1993): 170–74.

14. Hemila and Chalker, "Vitamin C for Preventing and Treating the Common Cold"; H. Hemila, "Vitamin C and Infections," *Nutrients* 9 (2017): 339–56; C. Jacobs, B. Hutton, T. Ng, R. Shorr, and M. Clemons, "Is There a Role for Oral or Intravenous Ascorbate (Vitamin C) in Treating Patients with Cancer? A Systematic Review," *Oncologist* 20 (2015): 210–23; M. A. Moser and O. K. Chun, "Vitamin C and Heart Health: A Review Based on Findings from Epidemiological Studies," *International Journal of Molecular Science* 17, 8 (2016): 1328, https://doi.org/10.3390/ijms17081328.

15. H. Gerster, "No Contribution of Ascorbic Acid to Renal Calcium Oxalate Stones," *Annals of Nutrition and Metabolism* 41 (1997): 269–82.

16. S. Wu, G. Wu, and H. Wu, "Hemolytic Jaundice Induced by Pharmacological Dose Ascorbic Acid in Glucose-6-Phosphate Dehydrogenase Deficiency," *Medicine* 97, no. 51 (2018): e13588.

APPENDIX: SELECTED FOOD SOURCES OF VITAMIN C

1. National Institutes of Health, Office of Dietary Supplements, "Vitamin C," https://ods.od.nih.gov/factsheets/VitaminC-HealthProfessional/.

BIBLIOGRAPHY

Aghajanian P., S. Hall, M. D. Wongworawat, and S. Mohan. "The Roles and Mechanisms of Action of Vitamin C in Bone: New Developments." *Journal of Bone and Mineral Research* 30 (2015): 1945–55.

Altman, Lawrence K. *Who Goes First? The Story of Self-Experimentation in Medicine.* New York: Random House, 1987.

Anderson, T. W. "Large-Scale Trials of Vitamin C." *Annals of the New York Academy of Sciences* 258 (1975): 498–504.

Anderson, T. W., G. H. Beaton, P. N. Corer, and L. Spero. "Winter Illness and Vitamin C: The Effect of Relatively Low Doses." *Canadian Medical Association Journal* 112 (1975): 823–26.

Anderson, T. W., D. B. W. Reid, and G. H. Beaton. "Vitamin C and the Common Cold: A Double-Blind Trial." *Canadian Medical Association Journal* 105 (1972): 503–8.

Anderson, T. W., G. Suranyi, and G. H. Beaton. "The Effect on Winter Illness of Large Doses of Vitamin C." *Canadian Medical Association Journal* 111 (1974): 31–36.

Anson, George. *A Voyage around the World.* London: Paean Books, 2011.

Aspin, R. K. "The Papers of Sir Thomas Barlow, BT, KVCO, FRS, PRCP (1845–1945)." *Medical History* 37 (1993): 333–40.

Baly, W. "On the Prevention of Scurvy in Prisoners, Pauper Lunatic Asylums, Etc." *London Medical Gazette* 1 (1843): 699–703.

Barlow, T. "On Cases Described as 'Acute Rickets' Which Are Probably a Combination of Scurvy and Rickets, the Scurvy Being an Essential, and the Rickets a Variable Element." *Medico-Chirurgical Transactions* 66 (1883): 159–220.

———. "The Bradshaw Lecture on Infantile Scurvy and Its Relation to Rickets." *British Medical Journal* 2, no.1767 (November 10, 1894): 1029–34.

Barrett, J. "Observations on Scurvy: As It Was Developed in Bath and Its Neighborhood, in the Spring of 1847." *Provincial Medical and Surgical Journal* 13, no. 6 (March 21, 1849): 148–53.

Bartley, W., H. A. Krebs, and J. R. P. O'Brien. "Vitamin C Requirements of Human Adults." Medical Research Council Special Report Series No. 280. London: Her Majesty's Stationery Office, 1953.

Bay, A. "Mori Ōgai Mori and the Beriberi Dispute." *East Asian Science, Technology and Society: An International Journal* 5 (2011): 573–79.

Beaglehole, J. C. *The Life of Captain James Cook*. Stanford, CA: Stanford University Press, 1974.

Berg, Jeremy M, John L. Tymoczko, and Lubert Stryer. *Biochemistry*. 5th ed. New York: W. H. Freeman, 2002.

Bjelakovic, G., D. Nikolova, R. G. Simonetti, and C. Gluud. "Antioxidant Supplements for Prevention of Gastrointestinal Cancers: A Systematic Review and Meta-Analysis." *Lancet* 364 (2004): 1219–28.

Blaine, Gibert. *Observations on the Diseases of Seamen*, 2nd ed. London: Joseph Cooper, 1789; repr. Boston: Gate ECCO, 2010.

Bollet, A. J. "Politics and Pellagra: The Epidemic of Pellagra in the U.S. in the Early Twentieth Century." *Yale Journal of Biology and Medicine* 65 (1992): 211–21.

Bown, Stephen R. *Scurvy: How a Surgeon, a Mariner and a Gentleman Solved the Greatest Medical Mystery of the Age of Sail*. New York: Thomas Dunne, 2003.

Boyle, P. P., and C. O. Grada. "Fertility Trends, Excess Mortality and the Great Irish Famine." *Demography* 23, no. 4 (1986): 543–62.

Brindan, Lynn, Alison Brading, and Tilli Taney, eds. *Women Physiologists. An Anniversary Celebration of their Contributions to British Physiology*. London: Portland Press, 1993.

Byron, Willian. *Cervantes: A Biography*. Garden City, NY: Doubleday, 1978.

Calahan, J., D. Howard, A. J. Almalki, M. P. Gupta, and A. I. Calderon. "Chemical Adulteration in Herbal Medicinal Products: A Review." *Planta Medica* 82 (2016): 505–15.

Camarena, V., and G. Wang. "The Epigenetic Role of Vitamin C in Health and Disease." *Cellular and Molecular Life Sciences* 73 (2016): 1645–58.

Carpenter, Kenneth J. *The History of Scurvy and Vitamin C*. Cambridge: Cambridge University Press, 1986.

———. Review of *Free Radical: Albert Szent-Gyorgyi and the Battle over Vitamin C*, by Ralph W. Moss. *Journal of Nutrition* 118 (1988): 1422–23.

———. *Beriberi, White Rice and Vitamin B*. Berkeley: University of California Press, 2000.

Carr, C. C., and S. Maggini. "Vitamin C and Immune Function." *Nutrients* 9 (2017): 1211, https://doi.org/10.3390/nu9111211.

Carter, K. C. "The Germ Theory, Beriberi, and the Deficiency Theory of Disease." *Medical History* 21 (1977): 119–36.

Chalmers, T. C. "Effects of Ascorbic Acid on the Common Cold." *American Journal of Medicine* 58 (1975): 532–36.

Chappell, Vere C., ed. *The Philosophy of David Hume*. New York: Modern Library, 1963.

Chatterjee, I. B. "Evolution and the Biosynthesis of Ascorbic Acid." *Science* 182 (1973): 1271–72.

Chatterjee, I. B., A. K. Mujumder, B. K. Nandi, and N. Subramanian. "Synthesis and Some Major Functions of Vitamin C in Animals." *Annals of the New York Academy of Sciences* 258 (1975): 24–47.

Chick, H., E. J. Dalyell, M. Hume, H. M. M. Mackay, and H. Henderson Smith. "The Etiology of Rickets in Infants." *Lancet* 2 (1922): 7–11.

Chick, H., E. M. Hume, R. F. Skelton, and A. Henderson Smith. "The Relative Content of Antiscorbutic Principle in Limes and Lemons." *Lancet* (November 30, 1918): 735–38.

Chick, Harriette, Margaret Hume, and Marjorie Macfarland, *War on Disease: A History of the Lister Institute*. London: Andre Deutsch, 1971.

Christison, R. "On Scurvy: Account of Scurvy as It Has Lately Appeared in Edinburgh, and of an Epidemic of It among Railway Labourers in the Surrounding County." *Monthly Journal of Medical Science* 13, no. 74 (1847): 1–22.

Constantini, N. W., G. Dubnov-Raz, B. Eyal, E. M. Berry, A. H. Cohen, and H. Hemila. "The Effects of Vitamin C on Upper Respiratory Infections in Adolescent Swimmers: A Randomized Trial." *European Journal of Pediatrics* 170 (2011): 59–63.

Cook, G. C. "Scurvy in the British Mercantile Marine in the 19th Century, and the Contribution of the Seaman's Hospital Society." *Postgraduate Medical Journal* 80 (2004): 224–29.

Cook, J. "The Methods Taken for Preserving the Health of the Crew of His Majesty's Ship the *Resolution* during Her Late Voyage around the World." *Philosophical Transactions of the Royal Society of London* 66 (1776): 402–6.

Cooper, E. A. "On the Protective and Curative Properties of Certain Foodstuffs against Polyneuritis Induced in Birds by a Diet of Polished Rice." *Journal of Hygiene* 12 (1912): 436–62.

———. "The Nutritional Importance of the Presence in Dietaries of Minute Amounts of Certain Accessory Substances." *British Medical Journal* 1, no. 2727 (1913): 722–24.

———. "On the Protective and Curative Properties of Certain Foodstuffs against Polyneuritis Induced in Birds by a Diet of Polished Rice." *Journal of Hygiene* 14 (1914): 12–22.

Coulehan, J. L., K. S. Reisinger, K. D. Rogers, and D. W. Bradley. "Vitamin C Prophylaxis in a Boarding School." *New England Journal of Medicine* 290 (1974): 6–10.

Cowan, D. W., H. S. Diehl, and A. B. Baker. "Vitamins for the Prevention of Colds." *JAMA* 120 (1942): 1268–71.

Crandon, J. H., and C. C. Lund. "Vitamin C Deficiency in an Otherwise Normal Adult." *New England Journal of Medicine* 222 (1940): 748–52.

Crandon, J. H., C. C. Lund, and D. B. Dill. "Experimental Human Scurvy." *New England Journal of Medicine* 223 (1940): 353–69.

Crowley, Roger. *City of Fortune: How Venice Ruled the Seas*. New York: Random House, 2011.

———. *Conquerors: How Portugal Forged the First Global Empire*. New York: Random House, 2015.

Cuppage, Francis E. *James Cook and the Conquest of Scurvy*. Westport, CT: Greenwood Press, 1994.

Dalrymple, William. *The Anarchy*. New York: Bloomsbury, 2019.

De Vreese, L. "Causal (Mis)understanding and the Search for Scientific Explanations: A Case Study from the History of Medicine." *Studies in History and Philosophy of Biological and Biomedical Sciences* 39 (2008): 14–24.

Drouin, G., J. R. Godin, and B. Page. "The Genetics of Vitamin C Loss in Vertebrates." *Current Genomics* 12 (2011): 371–78.

Drummond, J. C., and A. Wilbraham. "William Stark, M.D." *Lancet* 226 (1935): 459–62.

Dugan, James. *The Great Mutiny*. New York: G. P. Putnam's Sons, 1965.

Dunn, W. A., G. Rettura, E. Seifter, S. England, "Carnitine Biosynthesis from γ-Butyrobetataine and from Exogenous Protein-bound 6-N-Trimethyl-L-lysine by the Perfused Guinea Pig Liver." *Journal of Biological Chemistry* 259 (1984): 10764–70.

Dykes, M. H. M., and P. Meier. "Ascorbic Acid and the Common Cold. Evaluation of Its Efficacy and Toxicity." *JAMA* 231 (1975): 1073–79.

Eban, Katherine. *Bottle of Lies*. New York: Ecco Press, 2019.

Eijkman, C. "Antineuitic Vitamin and Beriberi." Nobel Lecture (1929). NobelPrize.org www.nobelprize.org/prizes/medicine/1929/eijkman /lecture/.

Eisenberg, D. M., R. C. Kessler, C. Foster, F. E. Norlock, D. R. Calkins, and T. L. Delbanco. "Unconventional Medicine in the United States." *New England Journal of Medicine* 328 (1993): 246–52.

England, S., and S. Seifter. "The Biochemical Functions of Ascorbic Acid." *Annual Review of Nutrition* 6 (1986): 365–406.

Fara, Patricia. *A Lab of One's Own: Science and Suffrage in the First World War*. Oxford: Oxford University Press, 2018.

Fernie, A. R., and T. Tohge. "Ascorbate Biosynthesis. A Cross-Kingdom History." *eLife* 4 (2015): e07527.

Foster, William, ed. *The Voyages of Sir James Lancaster to Brasil and the East Indies 1591–1603*. London: Hakluyt Society, 1940.

Funk, C. "On the Chemical Nature of the Substance Which Cures Polyneuritis in Birds Induced by a Diet of Polished Rice." *Journal of Physiology* 43 (1911): 395–400.

Gahche, J. J., R. I. Bailey, N. Potischman, and J. T. Dwyer. "Dietary Supplement Use Was Very High among Older Adults in the United States in 2011–2014." *Journal of Nutrition* 147 (2017): 1968–76.

Geller, A. I., N. Shehab, N. J. Weidle, M. C. Lovegrove, B. J. Wolpert, B. B. Timbo, R. P. Mozersky, and D. S. Budnitz. "Emergency Department Visits for Adverse Events Related to Dietary Supplements." *New England Journal of Medicine* 373 (2015): 1531–40.

Gest, N., H. Gaitier, and R. Stevens. "Ascorbate as Seen through Plant Evolution: The Rise of a Successful Molecule?" *Journal of Experimental Botany* 64 (2013): 33–53.

Gorman, Sara E., and Jack M Gorman. *Denying to the Grave*. New York: Oxford University Press, 2017.

Grijns, G. "Over Polyneuritis Gallinarum." *Geneeskundig Tijdschrift voor Nererlandsch-Indië* 41 (1901): 3–110. Published in English in Grijns, G. *Researches on Vitamins 1900–1911*. Gorinchem: J. Noorduyn en Zoon N.V., 1935, 1–108.

———. *Researches on Vitamins 1900–1911*. Gorinchem: J. Noorduyn en Zoon N.V., 1935.

Hager, Thomas. *Force of Nature: The Life of Linus Pauling*. New York: Simon and Schuster, 1995.

Harden, A., and S. S. Zilva. "Accessory Factors in the Nutrition of the Rat." *Biochemical Journal* 12 (1918): 408–15.

———. "The Antiscorbutic Factor in Lemon Juice." *Biochemical Journal* 12 (1918): 259–69.

Harrison, M. "Scurvy on Sea and Land: Political Economy and Natural History, c.1780–c.1850." *Journal for Maritime Research* 15 (2013): 7–25.

Harvie, David I. *Limeys: The True Story of One Man's War against Ignorance, the Establishment and the Deadly Scurvy*. Stroud, UK: Sutton, 2002.

Hasselhot, S., P. Tveden-Nyborg, and J. Lykkesfeldt. "Distribution of Vitamin C Is Tissue Specific with Early Saturation of the Brain and Adrenal Glands Following Differential Oral Dose Regimens in Guinea Pigs." *British Journal of Nutrition* 113 (1915): 1539–49.

Hawkins, Richard. *The Observations of Sir Richard Hawkins, Knight, in His Voyage into the South Sea in the Year 1593*. San Francisco: Elibron Classics, 2005. First published 1847 by the Hakluyt Society (London).

Haworth, W. N. "The Structure of Carbohydrates and of Vitamin C." Nobel Lecture (1937). NobelPrize.org. www.nobelprize.org/prizes/chemistry/1937/haworth/lecture/.

Hemila, H. "Vitamin C and Infections." *Nutrients* 9 (2017): 339–56.

Hemila, H., and E. Chalker. "Vitamin C for Preventing and Treating the Common Cold." *Cochrane Database of Systematic Reviews* CD000980 (2013).

Herman, Arthur. *How the Scots Invented the Modern World.* New York: Crown, 2001.

Hess, A. F., and M. Fish. "Infantile Scurvy: The Blood, the Blood Vessels and the Diet." *American Journal of Diseases of Children* 8 (1914): 385–405.

Hess, Alfred F. *Scurvy: Past and Present.* Philadelphia: Lippincott, 1920. Available at http://chla.library.cornell.edu.

Hirschmann, J. V., and G. J. Raugi. "Adult Scurvy." *Journal of the American Academy of Dermatology* 41 (1999): 895–906.

Hodges, R. E., E. M. Baker, J. Hood, H. E. Sauberlich, and S. E. March. "Experimental Scurvy in Man." *American Journal of Clinical Nutrition* 22 (1969): 535–48.

Hodges, R. E., J. Hood, J. E. Canham, H. E. Sauberlich, and E. M. Baker. "Clinical Manifestations of Ascorbic Acid Deficiency in Man." *American Journal of Clinical Nutrition* 24 (1971): 432–43.

Holst, A. "Experimental Studies Relating to 'Ship Beri-Beri' and Scurvy." *Journal of Hygiene* 7 (1907): 619–33.

Holst, A., and T. Frølich. "Experimental Studies Relating to Ship Beri-Beri and Scurvy." *Journal of Hygiene* 7 (1907): 634–71.

Hopkins, F. G. "The Analyst and the Medical Man." *Analyst* 31 (1906): 385–404.

———. "Feeding Experiments Illustrating the Importance of Accessory Factors in Normal Dietaries." *Journal of Physiology* 44 (1912): 425–60.

———. "The Earlier History of Vitamin Research." Nobel Lecture (1929). NobelPrize.org. www.nobelprize.org/prizes/medicine/1929/hopkins/lecture/.

Huntford, Roland. *The Last Place on Earth.* New York: Modern Library, 1999.

Hurley, Dan. *Natural Causes: Death, Lies, and Politics in America's Vitamin and Herbal Supplement Industry.* New York: Broadway Books, 2006.

Hutchinson, Robert. *The Spanish Armada*. New York: Thomas Dunne, 2013.

Institute of Medicine, Food and Nutrition Board. *Dietary Reference Intakes for Vitamin C, Vitamin E, Selenium and Carentenoids*. Washington, DC: National Academy Press, 2000.

Ivanov, B. N. "Role of Ascorbic Acid in Photosynthesis." *Biochemistry (Moscow)* 79 (2014): 282–89.

Johnson, Steven. *The Ghost Map*. New York: Riverhead, 2007.

Johnston, C. S., F. M. Steinberg, and R. B. Rucker. "Ascorbic Acid." In *Handbook of Vitamins*, ed. Janos Zempleni. 4th ed. Boca Raton, FL: CRC Press, 2007.

Jukes, T. H. "The Identification of Vitamin C, an Historical Summary." *Journal of Nutrition* 118 (1988): 1290–93.

Karlowski, T. R., T. C. Chalmers, L. D. Frenkel, A. Z. Kapikian, T. L. Lewis, and J. M. Lynch. "Ascorbic Acid for the Common Cold. A Prophylactic and Therapeutic Trial." *JAMA* 231 (1975): 1038–42.

King, C. G., and W. A. Waugh. "The Chemical Nature of Vitamin C." *Science* 75 (1932): 357–58.

Kinsman, R. A., and J. Hood. "Some Behavioral Effects of Ascorbic Acid Deficiency." *American Journal of Clinical Nutrition* 24 (1971): 455–64.

Koplan, J. P., J. Annest, P. M. Layde, and G. L. Rubin. "Nutrient Intake and Supplementation in the United States (NHANES II)." *American Journal of Public Health* 78 (1986): 287–89.

Kuhn, Thomas S. *The Structure of Scientific Revolutions*. 4th ed. Chicago: University of Chicago Press, 2012.

Kumar, M., and A. E. Axelrod. "Circulating Antibody Formation in Scorbutic Guinea Pigs." *Journal of Nutrition* 98 (1969): 411–44.

Lamb, Jonathan. *Scurvy: The Disease of Discovery*. Princeton, NJ: Princeton University Press, 2017.

Leach, R. D. "Sir Gilbert Blane, Bart, M.D. FRS (1749–1832)." *Annals of the Royal College of Surgeons of England* 62 (1980): 232–39.

Leshem, Y. "Plant Senescence Processes and Free Radicals." *Free Radical Biology and Medicine* 5 (1988): 39–49.

Levine, M., C. Conry-Cantelena, Y. Wang, R. W. Welch, P. W. Washko, K. R. Dhariwal, J. B. Park, A. Lazarev, J. F. Graumlich, J. King, and L. R. Cantilena. "Vitamin C Pharmacokinetics in Healthy Volunteers:

Evidence for a Recommended Daily Allowance." *Proceedings of the National Academies of Science* 93 (1996): 3704–9.

Levine, M., Y. Wang, S. J. Padayatty, and J. Morrow. "A New Recommended Dietary Allowance of Vitamin C for Healthy Women." *Proceedings of the National Academies of Science* 98 (2001): 9842–46.

Lind, James. *A Treatise on the Scurvy*, 3rd ed. London, 1772; repr. Birmingham, AL: Classics in Medicine Library, 1980.

Lloyd, Christopher. "The Introduction of Lemon Juice as a Cure for Scurvy." *Bulletin of the History of Medicine* 35 (1961): 123–32.

———. *The Health of Seamen*. London: Navy Records Society, 1965.

Lloyd, Christopher, and Jack L. S. Coulter. *Medicine and the Navy, 1200–1900*. Edinburgh: E. and S. Livingstone Ltd., 1961.

Lown, Bernard. *The Lost Art of Healing*. Boston: Houghton Mifflin, 1996.

Lunin, N. (1881): "Über die Bedeutung der anorganischen Salze für Ernährung des Thieres." *Zeitschrift für Physikalische Chemie* 5 (1881): 31.

Magiokinis, E., A. Beloukas, and A. Diamantis. "Scurvy: Past, Present and Future." *European Journal of Internal Medicine* 22 (2011): 147–52.

Maltz, A. "Casimer Funk, Nonconformist Nomenclature, and Networks Surrounding the Discovery of Vitamins." *Journal of Nutrition* 143 (2013). 1013–20.

Mandl, J., A. Szarka, and G. Banhegyi. "Vitamin C: Update on Physiology and Pharmacology." *British Journal of Pharmacology* 157 (2009): 1097–1110.

Marcus, D. M. "Dietary Supplements: What's in a Name? What's in the Bottle?" *Drug Testing Analysis* 8 (2015): 410–12.

Markham, Sir Clements R., ed. *The Voyages of Sir James Lancaster, KT, to the East Indies*. Miami: Hard Press, 2014.

McCollum, E. V., and M. Davis. "The Necessity of Certain Lipins in the Diet during Growth." *Journal of Biological Chemistry* 15 (1913): 167–75.

McCollum, Elmer V. *A History of Nutrition*. Boston: Houghton Mifflin, 1957.

Medical Research Committee. *Report on the Present State of Knowledge Concerning Accessory Food Factors (Vitamines)*. Special Reports Series No. 20. London: His Majesty's Stationery Office, 1919.

Medical Research Council. *Vitamins: A Survey of Present Knowledge.* Medical Research Council Special Reports Series No. 167. London: His Majesty's Stationery Office, 1932.

Moss, Ralph W. *Free Radical: Albert Szent-Gyorgyi and the Battle over Vitamin C.* New York: Paragon House, 1988.

Murad, S., D. Grove, K. A. Lindberg, G. Reynolds, A. Sivarajah, and S. R. Pinnell. "Regulation of Collagen Synthesis by Ascorbic Acid." *Proceedings of the National Academies of Science* 78 (1981): 2879–82.

Murphy, S. P., D. Rose, M. Hudes, and F. E. Viteri. "Demographic and Economic Factors Associated with Dietary Quality for Adults in the 1987–88 Nationwide Food Consumption Survey." *Journal of the American Dietetic Association* 92 (1992): 1352–57.

Nandi, A., K. Mukhopadhyay, M. K. Ghosh, D. J. Chattopadhyay, and I. B. Chatterjee. "Evolutionary Significance of Vitamin C Biosynthesis in Terrestrial Vertebrates." *Free Radical Biology and Medicine* 22 (1997): 1047–54.

Nytko, K. J., N. Maeda, P. Schafli, P. Spielman, R. H. Wengler, and D. P. Stiehl. "Vitamin C Is Dispensable for Oxygen Sensing in Vivo." *Blood* 117 (2010): 5485–93.

Offit, Paul A. *Do You Believe in Magic? The Sense and Nonsense of Alternative Medicine.* New York: HarperCollins, 2013.

Osborne, T. B., and L. B. Mendel. "Feeding Experiments with Artificial Food-Substances." Carnegie Institution Publication No. 156 (1911).

———. "The Relation of Growth to the Chemical Constituents of the Diet." *Journal of Biological Chemistry* 15 (1913): 311–26.

Parrow, N. L., J. A. Leshin, and M. Levine. "Parenteral Ascorbate as a Cancer Therapeutic: A Reassessment Based on Pharmacokinetics." *Antioxidants and Redox Signaling* 19 (2013): 2141–56.

Pauling, Linus. "The Nature of the Chemical Bond: Application of Results Obtained from the Quantum Mechanics and from a Theory of Paramagnetic Susceptibility to the Structure of Molecules." *Journal of American Chemical Society* 53 (1931): 1367–1400.

———. *The Nature of the Chemical Bond.* Ithaca, NY: Cornell University Press, 1960.

———. "Orthomolecular Psychiatry." *Science* 160 (1968): 265–71.

———. "The Significance of the Evidence about Ascorbic Acid and the Common Cold." *Proceedings of the National Academies of Science* 68 (1971): 2678–81.

———. *Vitamin C, the Common Cold and the Flu.* San Francisco: W. H. Freeman, 1976.

Pauling, L., and R. B. Corey. "Atomic Coordinates and Structure Factors for Two Helical Configurations of Polypeptide Chains." *Proceedings of the National Academies of Science* 37 (1951): 235–40.

———. "Configurations of Polypeptide Chains with Favored Orientations around Single Bonds: Two New Pleated Sheets." *Proceedings of the National Academies of Science* 37 (1951): 729–40.

———. "The Pleated Sheet, a New Layer Configuration of Polypeptide Chains." *Proceedings of the National Academies of Science* 37 (1951): 251–56.

———. "The Polypeptide-Chain Configuration in Hemoglobin and Other Globular Proteins." *Proceedings of the National Academies of Science* 37 (1951): 282–85.

———. "The Structure of Feather Rachis Keratin." *Proceedings of the National Academies of Science* 37 (1951): 256–61.

———. "The Structure of Fibrous Proteins of the Collagen-Gelatin Group." *Proceedings of the National Academies of Science* 37 (1951): 272–81.

———. "The Structure of Hair, Muscle and Related Proteins." *Proceedings of the National Academies of Science* 37 (1951): 261–71.

———. "The Structure of Synthetic Polypeptides." *Proceedings of the National Academies of Science* 37 (1951): 241–50.

———. "A Proposed Structure for the Nucleic Acids." *Proceedings of the National Academies of Science* 39 (1953): 84–97.

Pauling, L., R. B. Corey, and H. R. Branson. "The Structure of Proteins: Two Hydrogen-Bonded Helical Configurations of the Polypeptide Chain." *Proceedings of the National Academies of Science* 37 (1951): 205–11.

Pauling, L., H. A. Itano, S. J. Singer, and I. C. Wells. "Sickle Cell Anemia: A Molecular Disease." *Science* 110 (1949): 543–48.

Pemberton, J. "Medical Experiments Carried out in Sheffield on Conscientious Objectors to Military Service during the 1939–45 War." *International Journal of Epidemiology* 35 (2006): 556–58.

Penn-Barwell, J. G. "Sir Gilbert Blane FRS: The Man and His Legacy." *Journal of the Royal Naval Medical Service* 102 (2016): 61–66.

Peters, E. M., J. M. Goetzsche, B. Grobbelaar, and T. D. Noakes. "Vitamin C Supplementation Reduces the Incidence of Postrace Symptoms of Upper-Respiratory-Tract Infection in Ultramarathon Runners." *American Journal of Clinical Nutrition* 57 (1993): 170–74.

Pijoan, M., and E. L. Lozner. "Vitamin C Economy in the Human Subject." *Bulletin of the Johns Hopkins Hospital* 75 (1944): 303–14.

Piro, A., G. Tagarelli, P. Lagonia, A. Tagarelli, and A. Quattrone. "Casimer Funk: His Discovery of the Vitamins and Their Deficiency Disorders." *Annals of Nutrition and Metabolism* 57 (2010): 85–88.

Public Health England. *Government Dietary Recommendations*. London: The Stationery Office, 2016, https://assets.publishing.service.gov.uk /government/uploads/system/uploads/attachment_data/file/618167 /government_dietary_recommendations.pdf.

Rankin, A., and J. Rivest. "Medicine, Monopoly, and the Premodern State—Early Clinical Trials." *New England Journal of Medicine* 375 (2016): 106–9.

Ravenstein, Ernst G., ed. *A Journal of the First Voyage of Vasco da Gama, 1497–99*. New York: Burt Franklin, 2017. First published in 1898 by the Hakluyt Society (London).

Ritchie, C. "Contributions to the Pathology and Treatment of the Scorbutus, Which Is at Present Prevalent in Various Parts of Scotland." *Monthly Journal of Medical Science* 2, no. 13 (1847): 38–49.

Ritzel, G. "Critical Evaluation of the Prophylactic and Therapeutic Properties of Vitamin C with Respect to the Common Cold." *Helvetica Medica Acta* 28 (1961): 63–68.

Roddis, Louis H. *James Lind: Founder of Nautical Medicine*. New York: Henry Schuman, 1950.

Sabiston, B. H., and M. W. Radonski. "Health Problems and Vitamin C in Canadian Northern Military Operations." Defense and Civil Institute of Environmental Medicine Report No. 74-R-1012 (1974), www.mv.helsinki.fi/home/hemila/CC/Sabiston_1974_ch.pdf.

Semba, R. D. "The Discovery of the Vitamins." *International Journal of Vitamin Nutrition Research* 82 (2012): 310–15.

Shah, S. K., F. G. Miller, D. C. Darton, D. Duenas, C. Emerson, H. Fernandez Lynch, E. Jamrozik, N. S. Jecker, D. Kamuya, M. Kapulu, J. Kimmelman, D. Mackay, M. J. Memoli, S. C. Murphy, R. Palacios, T. L. Richie, M. Roestenberg, A. Saxena, K. Saylor, M. J. Selgelid, V. Vaswani, and A. Rid, "Ethics of Controlled Human Infection to Address COVID-19." *Science* 368 (2020): 832–34.

Smirnoff, N. "Ascorbic Acid Metabolism and Function: A Comparison of Plants and Animals." *Free Radical Biology and Medicine* 122 (2018): 116–29.

Smirnoff, N., and G. L. Wheeler. "Ascorbic Acid in Plants: Biosynthesis and Function." *Critical Reviews in Biochemistry and Molecular Biology* 35 (2000): 291–414.

Smith, A. H. "A Historical Inquiry into the Efficacy of Lime Juice for the Prevention and Cure of Scurvy." *Journal of the Royal Army Medical Corps* 1 (1919): 93–116; 188–208.

Stare, F. J., and I. M. Stare. "Charles Glen King, 1896–1988." *Journal of Nutrition* 118 (1988): 1272–77.

Stewart, C. P., and Douglas Guthrie, eds. *Lind's Treatise on Scurvy.* Edinburgh: Edinburgh University Press, 1953.

Stockman, R. "James Lind and Scurvy." *Edinburgh Medical Journal* 33 (1926): 329–50.

Stubbs, B. J. "Captain Cook's Beer: The Antiscorbutic Use of Malt and Beer in Late 18th Century Sea Voyages." *Asia Pacific Journal of Clinical Nutrition* 13 (2003): 129–37.

Svirbely, J. L., and A. Szent-Györgyi. "The Chemical Nature of Vitamin C." *Biochemical Journal* 26 (1932): 865–70.

———. "Hexuronic Acid as the Antiscorbutic Factor." *Nature* 129 (1932): 576.

Szent-Györgyi, A. "Observations on the Function of Peroxidase Systems and the Chemistry of the Adrenal Cortex: Description of a New Carbohydrate Deriviative." *Biochemical Journal* 22 (1928): 1387–410.

———. "On the Mechanism of Biological Oxidation and the Function of the Suprarenal Gland." *Science* 72 (1930): 125–26.

———. "Oxidation, Energy Transfer and Vitamins." Nobel Lecture (1937). NobelPrize.org. www.nobelprize.org/prizes/medicine/1937/szent-gyorgyi/lecture.

———. "Lost in the Twentieth Century." *Annual Review of Biochemistry* 32 (1963): 1–15.

Thomas, W. R., and P. G. Holt. "Vitamin C and Immunity: An Assessment of the Evidence." *Clinical and Experimental Immunology* 32 (1978): 370–79.

Tucker, J., T. Fischer, L. Upjohn, D. Mazzera, and M. Kumar. "Unapproved Pharmaceutical Ingredients Included in Dietary Supplements Associated with U.S. Food and Drug Administration Warnings." *JAMA Network Open* 1, no. 6 (2018): e183337.

Washko, P. W., Y. Wang, and M. Levine. "Ascorbic Acid Recycling in Human Neutrophils." *Journal of Biological Chemistry* 268 (1993): 15531–35.

Waugh, W. A., and C. G. King. "Isolation and Identification of Vitamin C." *Journal of Biological Chemistry* 97 (1932): 325–31.

Wharton, M. "Sir Gilbert Blane Bt (1749–1834)." *Annals of the Royal College of Surgeons of England* 66 (1984): 375–76.

Wilson, C. W. M., and H. S. Loh. "Common Cold and Vitamin C." *Lancet* 1 (1973): 638–41.

Wilson, E. A. "The Medical Aspect of the *Discovery*'s Voyage to the Antarctic." *British Medical Journal* 2 (1905): 77–80.

Wilson, Edward A. *Diary of the Discovery Expedition*. London: Blandford Press, 1966.

Wilson, L. G. "The Clinical Definition of Scurvy and the Discovery of Vitamin C." *Journal of the History of Medicine and Allied Sciences* 30 (1975): 40–60.

Woodham-Smith, Cecil. *The Great Hunger: Ireland 1845–1849*. New York: E. P. Dutton, 1980.

Young, V. R. "Evidence for a Recommended Daily Allowance for Vitamin C from Pharmacokinetics: A Comment and Analysis." *Proceedings of the National Academies of Science* 93 (1996): 14344–48.

Zilva, S. S. "Recent Progress in the Study of Experimental Scurvy." *Proceedings of the Royal Society of Medicine* 18 (1925): 1–9.

———. "Hexuronic Acid as the Antiscorbutic Factor." *Nature* 129 (1932): 943.

———. "The Isolation and Identification of Vitamin C." *Archives of Disease in Childhood* 10 (1935): 253–64.

ACKNOWLEDGMENTS

Michael Larsen was generous with his time and advice and instrumental in organizing two writing groups to which I belong. I am indebted to Michael and to the members of these groups: Richard Bailey, June Johnson, Harlan Lewin, Keh-Meh Lin, Barbara Michelman, Jane Pearson, Brigitte Schulze Pilibosian, Scott Salinger, Steve Sodokoff, Sydney Sauber, and Rolene Walker. Donna Ferriero, Jonathan Silberman, and Frank Sharp suffered through early drafts of the book and provided both encouragement and insight.

Cristen Iris provided expert editorial service, encouragement, and most important, wise counsel. She consistently went above and beyond the call of duty to help get this book published. Danielle Goodman edited an early version of the manuscript and Stacey Smekofske a late version. The defects remaining in the book are solely due to my own shortcomings and refusal to follow advice.

I also owe a debt to many librarians. The Mechanics Institute Library in San Francisco supports writers in a myriad of ways and was a great resource with a welcoming environment in which to work. The librarians of the Countway Library of Medicine at Harvard Medical School and the Bioscience, Natural Resources and Public

Health Library at the University of California, Berkeley, cheerfully helped me find obscure references. The medical library of the University of California, San Francisco, was a resource for many older books and journals.

I am grateful to Jake Bonar and the staff at Prometheus Books for bringing this book to fruition. They were highly professional and infinitely patient with a rookie author.

Anyone interested in vitamins must thank Professor Kenneth J. Carpenter, whose scholarly and encyclopedic history, *The History of Scurvy and Vitamin C*, is invaluable. It is the first place to go for those who want to delve more deeply into the history of vitamin C.

Of course, I can never sufficiently thank my wife, Susan Semonoff, whose love, encouragement, and forbearance were essential.

INDEX